GLOBAL SOLUTIONS
DEMANDING TOTAL ACCOUNTABILITY
FOR CLIMATE CHANGE

Tom Mullikin

Vox Populi Publishers, LLC
100 North Tryon Street, Suite 4700
Charlotte, NC 28202-4003

10 9 8 7 6 5 4 3 2 1

ISBN   978-0-9790178-5-8
978-0-9790178-6-5

CONTENTS

DEDICATION

I dedicate this book to my mother and father. My father, like so many in his time, stepped up to the great challenges of his generation. He left Purdue University and volunteered to spend most of the years of World War II in the mud and blood of the forward positions in Germany. He collected intelligence to prepare for the major battles which would decide the fate of the world. My efforts in this "quick read" are nothing more than my collection of "intel" and tolling of the bell for an impending battle that we must win. And to my mother who never allowed me to drop my head, never allowed me to quit, no matter what the odds. Thank you for insisting on persistence.

Finally, I would like to thank my team for making the publication of this book possible.

PREFACE

I remember well having dinner at the home of my boss during the 1988 Presidential primary contest and watching home movies of one of his early trips to Antarctica. At the time I was in my late twenties and considered myself—then as I do now—a student of government. What struck me most was this man's genuine interest in the environment and its impact on our legacy and viability as a country, and as a global population. The enthusiasm he demonstrated while regaling the small group who gathered in his den after dinner with details of the various flora and fauna of Antarctica and the reasons why this desolate location was important to us in the United States, left a lasting impression. His remarks burned in me an interest in the environment, and for seeking solutions to some of the larger and more complex environmental concerns of our day. At that time, Al Gore was a young United States Senator, and an aspirant seeking the Democratic nomination for the office of the President of the United States of America, and I was his South Carolina Campaign Manager.

I suppose like any professional I would prefer to tell you that my career is marked by my own strategic planning and individual accomplishment. The truth however is that the very best successes in my career have been due to the good fortune I have had to learn from some of the most dedicated and industrious men who have led our government and industry during the last 25 years. From Al Gore to John McCain, to Dan DiMicco, as a giant of domestic manufacturing, I have had the honor to learn from some of the best.

That said, my story is no different than many other Americans. I was born to a working class family in eastern North Carolina. We were instilled with a strong work ethic borne largely out of necessity. My ethic was further refined by the misfortune of being born with severely deformed feet, which earned me assurances that I would never walk. I add this only to give some meaning to the attitude within this brief work that no issue is too great to overcome. This is an attitude that I brought to overcoming my disability, and it is one that I bring to any complex environmental concern.

I am not a natural scientist, and I do not intend to introduce new theories to prove or disprove this phenomenon—global climate change. I do however intend to offer proof as to why we as concerned Americans should engage with this issue. I also hope to challenge you to look beyond the election year sloganeering and dig a little deeper into what may become the single biggest issue of our generation.

Global climate change threatens to have a severe impact on the fabric of the average American working family due to the potential negative environmental consequences of the issue, coupled with broader economic concerns. The only thing greater than the potential impact of global climate change has been the dearth of meaningful national leadership on this issue. While our nation and communities across our country suffer from the confluence of global economic and environment conditions, our federal officials seem more concerned with superficial election year sound bites.

Proposed solutions to date—such as the Kyoto Protocol—have exempted so-called developing countries such as China, India, and Brazil. Some scientists believe that these and other developing nations now produce more greenhouse gas emissions than the developed nations. While the United States remains the largest single contributor, it is predicted that Chinese will take this position from us in the next two years. Although we in the United States have a great deal we can do to make a real contribution, it is clear we can not solve this problem alone.

The treatises and debate thus far have done little but generate interest in moving U.S. manufacturing operations to countries that have scant environmental regulation and little to no environmental enforcement. This has contributed to the burgeoning economy of these "developing" economies (China's economy, for example, has grown four times faster than the economy of the United States in the past decade). Meanwhile, the United States has lost more than 3.5 million manufacturing jobs in the past eight years. This makes for a perfect storm—in which the U.S. working families have lost good jobs while we experience greater negative pressures on our environment as multi-national corporations move manufacturing to countries that refuse to enforce safe environmental standards.

In addition to this jaundiced and ineffective global debate, our own domestic consideration lacks courage and meaning. Much of the debate on emission reduction has centered on reduction of point source emissions for industry and power generators. These emissions have been vigorously regulated since passage of the Clean Air Act in 1967. While further regulation of these emissions might result in some level of reduction, together, they barely make up half of the overall domestic greenhouse gas emissions. Much of America's emission of greenhouse gases comes from sources which will require greater public involvement, such as residential use of electricity. These are the types of emissions where the great American people can make the most profound impact by expanded use of recycling, solar energy and other types of alternative energy.

I believe in America and our ability to meet even the greatest challenges. We have a track record of success, and we can meet this challenge as well. We can not meet it however, if our elected leaders choose political expediency over substantive debate. What we need is responsible education, total accountability, and a real, global solution for a global issue.

This book may not be well received by those who would prefer to bet on the stock market than on American ingenuity. It will not be for those who prefer unfettered greed over unlimited optimism for a better environment through fair trade.

GLOBAL SOLUTIONS
DEMANDING TOTAL ACCOUNTABILITY FOR CLIMATE CHANGE

Tom Mullikin

Chapter 1

In December 2005, my team traveled to Antarctica. The fifth largest of the seven continents, Antarctica is a beautiful but stark no-man's land that has attracted explorers and researchers since the early 1800's, when John Davis, an American sealer, became the first person to set foot upon the rocky shore. It is hard to describe exactly what it is like to sail through the choppy water, the boat straining under violent gusts of wind, navigating around chunks of ice as big as buildings, towards a shoreline shadowed by looming glaciers.

Antarctica is the coldest, driest and windiest continent. Despite the charm of the penguins and the serene majesty of the glaciers, the land and sea are treacherous and unpredictable. The water is freezing, the wind carries stinging ice crystals, and the land rises high, frequently causing altitude sickness in visitors. It is common that human eyes are blinded by the sun's reflection on the snow and ice. When cyclonic storms form over the ocean and move clockwise along the coast, their winds toss sea vessels like a child's toy.

As a bitter wind swept down the polar plateau and our ship swayed in the icy water, I questioned my decision to make the journey. Standing against the rail, I could hear the groan of the boat's hull and feel the sting of the spray from the frozen sea. We had come to an inherently dangerous place.

In 1773, the explorer James Cook crossed the Antarctic circle and circumnavigated Antarctica. While he didn't sight land, he did sight rock particles in the icebergs. He is often quoted as saying, "I make bold to declare that the world will derive no benefit from it." James Cook may have accurately sensed the desolation and danger of Antarctica, but he underestimated the significance of the icy continent. Antarctica is much more than a destination for thrill-seekers; it has become home to some of the most sophisticated and important research into the issue of global climate change.

The American public now firmly believes in the dangers of global climate change. Recent polling indicates that 70 percent of Americans believe climate change is occurring and is affecting catastrophic weather patterns, such as hurricanes. In response to this growing concern, governments are taking action: a group of states in the Northeast have formed the Regional Greenhouse Gas Initiative (RGGI); California adopted carbon caps, and has formed a regional pact with Washington, Oregon, New Mexico, and Arizona; every day another state announces a new legislative initiative to reign in greenhouse gas emissions. As the new Congress plans its agenda in Washington, and as politicians react to the chorus of voters' voices, we can expect to see even bolder plans proposed—both in scope and size. But the real dangers of climate change will not be solved by flashy programs rashly adopted as part of election year politics. Without effective and responsible global solutions to climate change, these efforts will go for naught—and, in fact, may backfire.

As it has become increasingly clear, we are living in a global environment—greenhouse gasses do not respect state or national boundaries. The actions of a single state, or even a group of states, is not sufficient to stop greenhouse emissions. For example, California's recently adopted plan aims for a reduction of 25 percent of California's greenhouse gas emissions. This reduction will amount to just a 0.13 percent reduction in global greenhouse gas emissions by 2020, assuming all things remain equal. But things will not remain equal—the costs imposed by such unilateral carbon cap programs will lead to a further exodus of industry from the United States. In fact, the U.S. Congressional Budget Office warned in September 2006 that "[i]f a domestic carbon pricing program significantly increased the prices of U.S. produced goods—and was not matched by efforts to reduce emissions in other countries—it could cause carbon intensive industries to relocate to countries without similar restrictions, diminishing the environmental benefits to a domestic program." These companies will set up shop in less regulated—and therefore less costly—nations. This migration results in a net loss for the environment. Our challenge, then, is to craft solutions that recognize the necessity of global reductions in greenhouse gas emissions, not just reductions in a few states, or even just one nation.

It is the rapid globalization and the confluence of trade and the environment which makes this issue even more complex. Environmental concerns have always been challenging, but in the past they have been more linear. Our country has taken a leadership position by enacting sweeping and enforceable environmental laws such as the Clean Air Act of 1967 (CAA), the Resource Conservation and Recovery Act of 1976 (RCRA), the Clean Water Act of 1977 (CWA), the Comprehensive Environmental Response, Compensation, and Liability Act of 1980 (CERCLA, or Superfund), and the Pollution Prevention Act of 1990

(PPA), just to name a few. These laws set high standards for the United States, but provided the basis for American leadership in the creation and adoption of international environmental conventions. But these laws also were limited to dealing with pollutants that primarily originated in and affected this country. We are now dealing with a frustrating set of challenges that require new thinking and problem solving matrices. The issues: greenhouse gases do not respect boundaries drawn by governments, and exploding economies in developing nations, which enforce scant or no environmental protection, offer safe havens for companies to pollute at will. As this scenario plays out, our environment is being besmirched at an alarming rate, while our economy and manufacturing base are being gutted, because multinational corporations are quickly moving their operations to these less restrictive foreign states. As James Gustave Speth, the Dean of the School of Forestry and Environmental Studies at Yale University and former advisor to Presidents Carter and Clinton explained in his book, *Red Sky at Morning: American and the Crisis of the Global Environment*:

> [O]ur national legislation was successful in curbing many environmental abuses domestically . . . [but] more of the same will not get us where we want to be in time to head off an era of unprecedented environmental decline. . . . The current system of international efforts to help the environment simply isn't working. The design makes sure it won't work, and the statistics keep getting worse. We need a new design, and to make that happen, civil society must take the helm.[1]

This is the new challenge; the new paradigm for real global solutions.

The failures of our existing "non-global" solutions are already becoming apparent in Europe. Some European countries have seen a 50 percent increase in electricity costs, caused by the Kyoto-imposed carbon trading system. These early adopters have already experienced the dangers of a failed, non-global solution. Kyoto has been labeled an "environmental and economic failure" by experts as signatories fail to meet their emissions requirements, and "developing" countries—exempted from Kyoto's caps—reap the benefits of fleeing industry. The environmental "backfire" of carbon cap and trade programs could prove to be

5

devastating. For example, industry in "developing" countries—such as China—emits on average five times more greenhouse gasses than industry in the United States. As industry "leaks" from efficient and responsible developed nations under the strain of increased regulatory costs, this backfire will unfortunately result in increased worldwide greenhouse gas emissions.

Unilateral caps by governments in the United States threaten to force our businesses to seek nations where production costs are lower, or simply go out of business. And yet consumers' demand for goods will not subside simply because our manufacturers are gone. Instead, manufacturers in other nations—nations such as China—will pick up the slack. And these nations are definitely not environmentally responsible.

The International Energy Agency recently reported that China will become the world's worst emitter of carbon dioxide by 2009, at the latest. Meanwhile, China's economic growth has been explosive. China's economy has grown 142 percent in the past decade—four times faster than the U.S. To fuel this industrial expansion, China's energy consumption has increased dramatically. China's oil consumption has doubled in the past 20 years, and China is now the world's second largest consumer of oil. China's consumption of coal increases approximately 10 percent per year, and it is now the world's largest producer and one of the top importers of coal. Automakers anticipate that China will account for 18 percent of global car sales between 2002 and 2012, even further increasing their emissions from fossil fuels (in comparison, Americans will account for only 11 percent of global car sales in the same period).

This economic expansion has come at a high cost to China's environment. The World Health Organization found that China had seven of the world's 10 most polluted cities. Acid rain falls in about one-third of the nation. By 2025, scientists estimate that China will emit more carbon and sulfur dioxide than the U.S., Japan, and Canada, combined. More than 53 percent of the world's mercury emissions come from Asia. The Environmental Protection Agency has traced high levels of mercury in the United States back to Chinese sources, such as unfiltered coal-fired electric plants. China has roughly 2,000 coal-fired power plants, which account for 76 percent of China's energy supply. The Chinese government plans to double generating capacity in the next 15 years. Most of these new plants also will rely on coal combustion.

If the United States acts only to further reduce its own emissions it will simply exacerbate the pollution problem by fueling further economic expansion—and environmental abuse—in nations like China. The U.S. has an opportunity to lead the world in seeking a global solution to a global problem; a solution that holds all parties to the same standard; that holds all parties accountable for our global environment. Pollution does not respect national borders, and solutions that adhere to the fiction that pollution can be solved by one city, one state, or one nation, are doomed to failure. In fact, such "solutions" will only make the problem worse.

The United States must demand total accountability in regards to global greenhouse gas emission reductions. Anything short will be more election year politics, a failed environmental solution, and an economic threat. The United States has already lost almost 4 million manufacturing jobs in the past eight years to countries such as China due to unfair trade practices, such as dumping and currency manipulation. Aggressively enforcing free trade laws—such as by demanding China fairly value their currency—would produce positive environmental benefits by providing a level playing field for American industry, the most environmentally efficient industry in the world, to compete and grow.

A couple of key statistics provide insight into a real solution for global greenhouse gas emissions. First, as noted, China, where there are no recognizable environmental standards, will overtake the United States as the world's largest emitter of greenhouse gases by the time the next United States President is sworn into office (and emissions from so-called "developing nations" will exceed emissions from all developed nations). In fact, recent reports suggest that China, which has a long history of underreporting their emissions, may already be the world's largest greenhouse gas emitter. Second, American industrial point sources of pollution constitute only 4.5 percent of the global total of greenhouse gas emissions. Within these factual parameters, we can craft a global solution: creating meaningful and enforceable international standards on all countries, and seeking meaningful reductions in all sectors of emissions.

A true global solution must account for the potential "migration" of emissions from regulated to unregulated nations. Regulations that increase costs on industry without taking account of the movement of industry to unregulated nations threaten to exacerbate the overall problem. Unfortunately, many of the cap and trade programs that have been proposed in the United States fall short in this respect. An alternative solution the United States should consider is the implementation a carbon based greenhouse gas (GHG) excise tax.

Excise taxes are taxes levied upon specific goods, unlike sales taxes or value added taxes, which are applied across a range of items. Taxes on cigarettes, alcohol and gasoline are examples of excise taxes. These types of taxes are often used to discourage a certain type of behavior. Excise taxes have successfully been used to discourage environmental pollution. In 1988, in response to mounting international concern over the depletion of the ozone layer, the United States ratified and implemented the "Montreal Protocol on Substances that Deplete the Ozone Layer." As part of its effort to ensure that the United States met its obligation under the Montreal Protocol, Congress enacted an excise tax on ozone-depleting chemicals (ODCs). The tax required manufacturers, producers and importers to pay an excise tax for the sale or use of ODCs in the production process, thus increasing the price of ODCs worldwide and providing incentives for the development of alternative chemicals. This led to a significant reduction in the production of ozone-depleting chemicals and had a measurable effect on environmental quality.

A carbon-based GHG tax on domestic and imported products would be levied on products based on the amount of carbon emitted during the manufacturing process, and would create the proper incentives for change both in the United States and abroad. Such a tax would eliminate the threat of emissions "migration" and would use market incentives to encourage other nations, such as China, to reduce the carbon intensity of their products in order to compete with more responsible nations. It would also avoid one of the major pitfalls of a treaty; as Sophocles once opined: "No treaty is ever an impediment to a cheat."

The United States can also lead by encouraging and promoting more recycling to recapture invested energy, more green technology, and more clean energy. Recently, New Mexico Governor Bill Richardson's committee on climate change issued a study that presented a broad range of opportunities to reduce energy consumption while simultaneously increasing production of clean energy, all while creating jobs and economic opportunities within the state. Similarly, California recently passed the Million Solar Roofs bill, which provides incentives for homeowners to install solar electricity. Among other things, the Million

8

Solar Roofs Bill provides ten years of incentives to help residents install one million solar energy systems on homes and businesses throughout the state by 2018. The result will be 3,000 Megawatts of new solar energy which is roughly equivalent to 40 peaking power plants.

By creating incentives for green industry and clean energy to grow and thrive, we can positively affect the environment through economic stimulus, without risking an exodus of American industry to environmentally irresponsible nations such as China.

It is in this way we can both save ecosystems such as the one we explored in Antarctica, and the fragile environments of African nations, such as Namibia. At the same time, we can save our communities back home.

Chapter 2

The Drake Passage is one of the most stormy and turbulent regions of ocean in the world. It is a 400-mile wide passage between Antarctica and South America. The Antarctic circumpolar current, which flows unimpeded at this latitude, picks up speed as the current enters the Passage, which is the narrowest point between Antarctica and any other continent. Those who must sail through the Passage often encounter waves over 30 feet high. Our journey to Antarctica required a trip through these waters, and the Passage made sure to live up to its reputation.

I've spent decades making my way through troubled waters. I don't know how it started, exactly. Maybe it was the way I was raised. But we've put together a team that is interested in trying to help businesses and communities work together to deal with the global issue of climate change. Guess you could say we like a challenge.

Nevertheless, I enjoy looking for ways to solve problems that benefit everyone involved. Over the course of my career, I have held a number of senior legal, professional and governmental positions. I have served on state and congressional staffs and as an advisor and campaign manager to senior members of Congress and U.S. Presidential candidates from both major parties. During all this time the most important lesson I've learned, both in and out of my government work, is that by merely bringing people together and educating them one can bring about a positive solution to most any public policy issue.

Really, it's true.

The difficulty with most issues is that people take a position, draw a line in the sand, and do not move. This makes them feel empowered, when, in reality, their inability to consider their beliefs in a critical light often renders their voices ineffective. Sadly, I've seen people become so invested in their position that they end up destroying the very things they care about.

The philosopher Gerald Nosich writes, "Effective thinking requires us to pay attention to the complexity of things. It requires us to develop a tolerance for ambiguity and an acceptance of less-than-certain answers. It requires a commitment to seeing both sides of an issue and to trying to find out the truth, rather than merely trying to bolster our side."[2] I think this is an accurate description of how I try to look at the world around me. It is my team's goal to use reasoned, balanced thinking to solve problems. As attorneys in this field, it is also our goal to use law, policy and social devices to advocate and create win-win situations that bring people together in a way that affects real, lasting solutions to problems.

People often have a way of thinking that hinders their ability to search for a solution to a problem. The goal of my team is to be facilitators of mutually beneficial debate between businesses and communities and the politicians that serve them both. We want to help the citizen and the CEO, to give them the tools to speak to one another and achieve results.

Sometimes, this seems impossible to do. Businesses need to make money. This means that they need to run cost-effectively. Communities want things too: they want businesses to give back to the neighborhoods in which they exist, and they want to keep the land safe and clean. Often these goals seem mutually exclusive.

But they're not.

By enveloping the broader concerns of a community in which a corporation wishes to operate, businesses can seek out mutually beneficial solutions where corporations and their shareholders benefit financially, and at the same time provide excellent community environmental stewardship. This is the goal—mutual benefit. Education and awareness are absolutely essential for all stakeholders to benefit. This is particularly critical with global climate change. By raising awareness—educating all stakeholders on the real issues—we can meet the real and perceived challenges. It is critical that there be total accountability, not failed, partial solutions.

This lesson became most vivid to me in 1998, in South Central Los Angeles. At the time, I was Vice President of Public Affairs and Chief Counsel for Safety-Kleen Corporation. Safety-Kleen was proud to claim the title of North America's largest environmental services company. We had a transfer station for used solvents in South Central Los Angeles (a permitted hazardous waste treatment storage and disposal facility (TSDF)) that was operating in the heart of an area that had seen a lot of damage during the Rodney King protests of 1992. Thousands of people in Los Angeles, primarily young black and Latino males, were driven to violent protest when a mostly white jury acquitted four police officers who had been videotaped brutally beating Rodney King, a black man who had led them on a 110 to 115 MPH freeway chase. Furious at the seemingly racist leanings of the California legal system, protestors swarmed over the city of Los Angeles, vandalizing businesses and burning buildings. A few innocent bystanders were beaten. When the smoke cleared, it was discovered that 3,600 fires had been set, 50 to 60 people had been killed, and material damages totaled more than $800 million.[3] Six years later, South Central L.A., and specifically the community around the facility where I was working, had not recovered. There were bars on the windows and angry faces on the sidewalks. Businesses were suspicious of people in the community, and the community believed the businesses were part of a system that left them feeling oppressed and abused.

Needless to say, the community surrounding the TSDF was not happy to have a waste management facility in its backyard. Despite the fact that they existed side by side, the company and the community saw one another as threats. The company did not trust the people who lived nearby, and the citizens believed that the waste company was poisoning the land on which they lived. In an effort to overcome this mutual distrust, we reached out to the Concerned Citizens of South Central Los Angeles ("CCSCLA"), one of the state's oldest and most respected environmental justice organizations.

As Safety-Kleen's representative, I found myself meeting with Juanita Tate, the executive director of CCSCLA, a lifelong advocate for environmental and social responsibility, and a woman whom I remember to this day. She was one of the key leaders in her community, and she knew *everybody* in that community. A sharp, sensible woman, she lived in a dangerous neighborhood, where I was warned to stay in my cab when I first went to see her. I did get out of the cab, though, and I am grateful that I did so.

We understood the importance of engaging and involving the public. We wanted to cultivate a staff of dependable employees, not a neighborhood of protestors. So they came to the table with an open mind. And between Safety-Kleen and CCSCLA, the corporation and the community found a way to do more than just to co-exist.

After months of discussion, the dialogue between the two groups came to focus not only on the issues that put the company and the community at odds, but also on the ways in which they might help one another. Ultimately, both sides determined to reach some sort of proactive solution to the challenges they both faced.

We created PET: The Partnership for Environmental Training. The program was designed with two purposes in mind: a) to give local people an opportunity to benefit from the company's presence and b) to improve the community's negative opinion of the company. Specifically, the program provided young men and women in the South Central community with the requisite skills to seek and attain good-paying, high-tech jobs in the environmental services industry. The program itself consisted of 80 hours of instructional classes sponsored by Safety-Kleen at the company's South Central Los Angeles facility. Additionally, the company underwrote the cost of pre-employment medical exams and drug screenings; they paid for respirators, safety boots, work clothes, bus fares and all training materials.

A seven-member volunteer advisory council, the majority of which was appointed by the community group led by Juanita Tate, managed the program, and the council's three remaining seats were held by representatives from Safety-Kleen. In 1999, just one year after Safety-Kleen and CCSCLA began their talks, a total of 72 students graduated from the program.

I am proud to note that the advisory council was able to set up temporary employment for each program graduate.

Some graduates began working for Safety-Kleen, while others were placed in positions with other corporations—including some of Safety-Kleen's competitors. The program's goal—to ensure that the trainees mastered a new set of technical skills that they would have the opportunity to put to work—had been met.

PET succeeded because it was structured around a mutual understanding of the goals of the business and the community. It took into account the needs and interests of both parties. Before PET stepped in, the community struggled with an unemployment rate of approximately 52 percent, and citizens wanted jobs. In response, the program provided meaningful employment opportunities in an area where such opportunities were few. Correspondingly, the company needed educated, professionally trained employees. For months they had been busing in temporary workers from all over the state: workers who were not invested in maintaining a job so far from their homes. Temporary employment agencies in the area had been spending up to $1,500 per person to screen, train and otherwise prepare outside employees for work on the SCLA job site! PET provided the company with trained local employees to help it with its business.

Ultimately, PET enabled each group to benefit from working with the other. The community grew to feel so strongly supportive of the company that, when it came time for permit renewal in 1999, citizens and community employees went to the state capitol in Sacramento to publicly support the renewal.

New jobs for the citizens and committed, trained workers for the company were not the only fruits of the project. Other local businesses and organizations flourished once residents were able to find meaningful jobs. Finally, the City of Los Angeles benefited because PET graduates were equipped with new skills that enabled them to participate in household hazardous waste cleanups throughout the city. By helping to rid neighborhoods of dangerous materials, they were able to help create safer communities for California families and children.

Without PET and its efforts to work with both the citizenry and Safety-Kleen, both groups would have suffered. Later, I published the details of my experiences in the *UCLA Journal of Environmental Law and Policy*, and explained it this way:

> The conflicts between industries and their host communities are likely to continue to increase in severity until and unless innovative methods of relationship building are developed. If history is any indication, these antagonistic relationships will continue to spill over into the courts producing costly, time-consuming litigation that does not serve the environmental, community or business interests. This is especially true regarding citizen lawsuits in which community representatives feel they have been denied a voice in the process. Collaborative models, such as that defined by the Partnership for Environmental Training, have tremendous potential for building partnerships, defusing hostilities between stakeholders, and creating more win-win situations. [4]

The outcome of the situation in SCLA was affirmation that this program of grassroots activism could yield dividends.

Drawing on my experiences in South Central Los Angeles, I have taken up the work of encouraging citizens and industries to settle their differences without resorting to lawsuits. I became especially concerned with the Environmental Justice Movement of North Carolina and South Carolina, to which I brought attention in an article for the *UCLA Journal of Environmental Law and Policy* in 2002. The Environmental Justice Movement originated in 1982 and was predicated on the ability of citizens to seek "environmental equity" through Title VI of the Civil Rights Act. More specifically, the movement began with hundreds of African Americans who joined together to protest North Carolina's plan to bury PCBs in a proposed landfill in Warren County. Leaders of the United Church of Christ, the Southern Leadership Council, and Walter Fauntroy, the U.S. House delegate from the District of Columbia joined local citizens in protesting the proposed facility. The group argued that the site had been chosen for the wrong reasons: namely, because it was one of the poorest regions in the state and because it was home to the highest percentage of minorities of any North Carolina county.

Tensions escalated when 400 demonstrators were arrested; the situation attracted the attention of the national news media. As a consequence, Governor Jim Hunt declared that no new landfills would be built in Warren County. Nevertheless, the issue of environmental justice in the Carolinas remains a troubled one. As I argued in the *UCLA Journal of Environmental Law and Policy*:

> [L]ittle consensus exists as to the validity of charges that environmental threats are being forced upon communities of color because those areas do not possess the political or economic strength to resist. [. . .] After almost twenty years of study, this issue is still defined more by conflict than common ground. Environmental justice remains mired in a policy war of attrition that pits citizen activists against the business community. [. . .] Rather than forging a constructive dialogue that at least respects the concerns of all stakeholders, the debate is too often fueled by agitation, self-interest, politics, partisanship, and rhetoric.[5]

My goal has been and continues to be mutual benefit and reasoned policy. "Environmental public participation rights make good economic sense."[6] Public participation on issues that concern the use and preservation of the environment is necessary to industrial development. Sound economic growth can occur only when companies are able to inspire the support and involvement of the host communities in which they exist. Why is this? There is a critical relationship between the industrialization of an area and the need for the environmental regulation of that area. Because there will always be public concern for the protection of public air, land, and water, the public should have a voice in the regulation of those elements. I wrote in the *Georgetown International Environmental Law Review* in 2005:

> As States move through the stages of industrialization, the resulting environmental stressors intensify concern for the protection of the environment among a public that ultimately bears many of the externalities of development. Failure to recognize this phenomenon and to include the public as participants in environmental decision-making will have long-term negative consequences on industry. Therefore, public support is crucial to continued and sustainable development.[7]

In some ways, this idea is a relatively new one. Citizens of ancient Greece and Rome, as well as those of the Middle Ages in Europe, did not clearly understand the extent of the effect human production and waste could have on the land on which they lived. It wasn't until the industrialization of the nineteenth century that people really became aware of pollution problems. Only when unregulated industry produced serious levels of water and air pollution did governments begin to heed public calls for environmental regulation. Today, it is the rapid confluence of greenhouse gases from all of the world's population gives rise to global climate change.

It was with the first Public Health Act—passed in Great Britain in 1875—that government was given a responsibility to work towards the improvement of the water supply and to appropriately find ways to eliminate waste and refuse. Almost a century later, after World War II, conservation-based environmental movements convinced middle America to demand that government not only help to provide citizens with cleaner public spaces but that it regulate modern industries.

Simlarly, as the world has grown smaller, these same dynamics are playing out in the global environment.

Currently there is a growing global acceptance of the rights of the public to participate in the regulation of the environment. This can be seen specifically in the field of International Law. Public participation in environmental issues has been embraced internationally as a key ingredient to sustainable development. Many international aid agencies and private institutional investors (such as The United States Agency for International Development [USAID], the World Bank, and the Asian Development Bank) see public participation as key to the growth and development of a society. Indeed, they are more likely to give money to developers and host nations that involve the public in environmental decisions. Additionally, capital investors are more likely to invest in state businesses that implement plans that not only allow for but encourage public participation. The *Georgetown International Environmental Law Review* said,

> This "nexus" between public participation and successful economic growth is reflected in the historical experiences of developed and heavily industrialized States such as the United States and the States that make up the European Union. It is no coincidence that these States have created elaborate mechanisims for environmental public participation rights and have been the most successful at injecting those rights into broad, far-reaching environmental legislation. A study of this historical nexus demonstrates how public concern resulting from industrialization created the impetus for the current public participation schemes embedded in the environmental legislation of these developed nations.[8]

As other states experience their own economic, social, or industrial revolutions, it would behoove them to follow these examples and to secure environmental participatory rights for their citizens. In addition to gaining international favor by this move, transitional states may minimize the potential negative impacts of social and economic change on their populations. Often, when affected citizens are left out of the development process, transitional governments face protests, riots, and additional long-term expenses.

In 2003, my team traveled to Prague to speak on this issue at the the Sixth International Symposium & Exhibition on Environmental Contamination in Central and Eastern Europe and the Commonwealth of Independent States. We emphasized the legal requirements that ensure community participation in environmental protection and management. As stated previously, we believe that if government will recognize the public's views on environmental issues, that both citizens and countries will see progress; such cooperation can be profitable for both economies and for the state of a commonwealth. We also traveled to Moscow in 2003 to speak about how Russia might develop a waste management system (to manage the millions of tons of industrial solid waste and hazardous waste produced in that nation each year). In 2003, we proposed a plan to governmental officials of Spain in an effort to help them deal with the social and environmental consequences of the shipwrecked oil tanker *Prestige.*

Public involvement in environmental issues is closely linked to the development of a civil society. An awareness and respect of the world is linked to an awareness and respect of the people who live within that world. As the Dalai Lama writes, "Our ancient scriptures speak of the container and the contained. The world is the container, our house, and we are the contained, the contents of the container. From these simple facts we deduce a special relationship, because without the container, the contents cannot be contained. Without the contents, the container contains nothing. It's meaningless."[9] We exist in symbiosis with our planet.

The recognition of public involvement in various *international* environmental initiatives (by way of conventions, agreements, declarations and general international policy statements) is a relatively new phenomenon. Nevertheless, such recognition is growing. Public participation in environmental management can be seen as an inherent right of all people living in organized societies. Although different countries may have different ways of interpreting and implementing these rights, the recognition of the need to involve the public in debates about the world in which they live is intimately involved in development of governing principles.

A number of recent International Environmental Initiatives emphasize this theme: the Rio Declaration made at the United Nations Conference on Environment and Development,[10] the Basel Convention of

Transboundary Movements of Hazardous Wastes and their Disposal,[11] the Kyoto Protocol to the UN Framework Convention on Climate Change,[12] the Rotterdam Convention on Prior Informed Consent Procedure for Certain Hazardous Chemicals and Pesticides in International Trade,[13] the UNECE Convention on Access to Information,[14] Public Participation in Decision making and Access to Justice in Environmental Matters at the Aarhus Convention,[15] and the Johannesburg Declaration on Sustainable Development.[16] Perhaps most significant to me, is the Rio Declaration, made in 1992 at the United Nations Conference on Environment and Development (UNCED, also known as Earth Summit). It states:

> Environmental issues are best handled with the participation of all concerned citizens, at the relevant level. At the international level, each individual shall have appropriate access to information concerning the environment that is held by public authorities, including information on hazardous materials and activities in their communities, and the opportunity to participate in decision-making process.[17]

The declaration also notes:

> One of the fundamental prerequisites for the achievement of sustainable development is broad public participation in decision-making [. . .] This includes the need of individuals, groups and organizations to participate in environmental impact assessment procedures and to know about and participate in decisions, particularly those which potentially affect the communities in which they live and work.[18]

We remain particularly committed to the issue of public involvement—not just on a national, but also on an international level. We remain dedicated to providing citizens of all states with access to information, with participation in the processes of decision-making, and with access to justice. In turn, the recognition of the need for public involvement in international environmental initiatives is rooted in pluralistic political views. This way of thinking emphasizes the idea that it is not possible to determine externally a single optimal solution to any social problem. The overall goal of public participation is to build trust and credibility, and to keep emotions and human energy focused on significant issues and on feasible solutions to social problems. Public participation provides an opportunity for all interested parties to become informed and involved, and to influence program development and implementation. Active public participation provides a forum to identify and address concerns, thus reducing conflict.

In his recent book, *Earth in the Balance*, Al Gore opined that we should "seek fundamental changes in how we gather information about what is happening to the environment and organize a worldwide education program to promote a more complete understanding of the crisis. In the process we should actively search for ways to promote a new way of thinking."[19] Gore also suggests that we need "successful negotiation and resolution of an entirely new generation of international treaties and agreements aimed at protecting the environment. Just as democracy and market economies are important to successful efforts to restore the earth's ecological balance, so is a further extension of the rule of law."[20] In these passages, the former Vice President makes very important observations. First, we need international education and understanding. Through this heightened global awareness, we can craft meaningful global solutions that will allow for total accountability. Next, he suggests that we need a new generation of international treaties. This, too, is profoundly insightful. Interestingly, however, he insisted on the adoption of the Kyoto Protocols, a treaty that led the world's population away from a meaningful global resolution and towards a failed political gimmick: a treaty which effectively excludes more than half of the world's greenhouse gas polluters from any real reductions. So counterintuitive is this failed initiative as to draw into suspicion the underlying motives. Why have we debated the "merits" of a proposal which gives a pass to burgeoning economic engines such as China and India? Particularly given recent projections that China will soon surpass the United States as the largest contributor of greenhouse gas emissions? We need to be bold enough to consider meaningful solutions; solutions that will manifest only after international education and understanding.

Currently I lead the Government, Policy, and Regulatory Affairs Team for the Charlotte, North Carolina, law firm of Moore and Van Allen. My practice addresses matters of corporate compliance and legislative representation. Consequently my team and I are aware of the issues affecting businesses and communities in the United States. Our goal is to remain focused on orchestrating proactive solutions to the problems that drive economies and communities, and we feel that our public policy model can help solve the crises facing both the citizen and the CEO.

So why did we go to Antarctica? We traveled many miles to that place of bitter cold for a purpose: to see firsthand the evidence of global climate change. Despite the ice, the isolation, and the risk, we were determined to learn to speak knowledgeably, both to our clients and to our communities, about the issues impacting our world and our businesses. The only way to do this was to see for ourselves the deterioration of the Antarctic ice shelf and to speak personally with those who know that the air of the polar regions is changing.

It may seem like an unusual choice for an attorney from Camden, South Carolina. However, I'm just trying to keep on doing what I've done all along. We see a problem with the land, and we see a public that is aware of that problem. We see businesses in conflict with the public, and we hope that our team can bring the people and the businesses together in order to help them find real solutions that can benefit both groups.

We are not alone in this. Many businesses are starting to plan for a future that includes a new climate, including Nucor Steel, Dupont, General Electric, BP, AES, and Whirlpool, to name just a few. These companies and others are already changing the way they do business, and indeed, most are improving both their profits and their customer relations by going green.

We owe our communities and our clients the responsibility to take positions and find out more about issues that are going to affect them. There is no removing the impact of this issue in the long term from industry in the United States. The effects of a changing climate are substantial. They will need to be addressed—not only by scientists and governments, but also by businesses and communities. This is why my team and I have chosen to actively research global climate change. We know it will present both a challenge and an opportunity for American companies that choose to proceed proactively. We hope to help those businesses plan for the future.

Chapter 3

We organized our journey to Antarctica in November 2005. In an effort to understand fully the climate situation, we gathered researchers, environmental experts, and legal professionals to join the expedition. Perhaps most notably, we were assisted in our quest by renowned environmental expert Dr. Robert Thomas, Chair of the Department of Environmental Communications at Loyola University in New Orleans.

Once there, we learned a lot.

In addition to the tsunamis, disastrously powerful hurricanes, and drought-fueled fires that appear on the nightly news, one can see in Antarctica the ways in which the world is changing. In Antarctica, the ice rose up before us like a magnificent colossus, yet it also began to fall back into the darkness of the sea. The silver rivulets of water ran down the glaciers, threatening their disintegration or future collapse. As we sailed towards the shore, we spoke with Dr. Richard Cameron, glaciologist and former program manager at the National Science Foundation. He pointed out to us the ways in which the Larsen B ice shelf—the major ice shelf in Antarctica—is falling apart. Dr. Cameron's words to us were, to put it bluntly, shocking ones. He said: "It is something the glaciologists never really considered . . . well we've always considered that the ice shelves would slowly melt back, if you will. But the disastrous disintegration of this ice shelf is something [. . .] really unthought-of. They didn't think such a thing could happen."

But it has happened. Over the past fifty years, the Antarctic Peninsula has warmed. And the ice is breaking off from the shelf to melt—in clumps that are the size of small countries. Similar melting is occurring in the Artic—2007 was ushered in with the news that the Ayles Ice Shelf in the Canadian North had broken free in late 2005. The shelf was the size of 11,000 football fields. These rapidly disintegrating ice shelves are heightening public fears.

In both heat waves and hurricanes, the global climate seems to be rumbling ominously around us. Although scientists have long claimed that global climate change will affect our environment, the issue itself, more recently, has affected our culture. Polls show a dramatic rise in the American public's awareness of this issue. In a 2006 *TIME* / ABC News / Stanford University poll, 85 percent of those polled believe that global warming is happening. Furthermore, most respondents claimed that they want to see action taken in response to this phenomenon. Of those polled, 87 percent believe that the government should either encourage or require power plants to lower their emissions; 87 percent believe companies should be given tax breaks to develop alternate energy sources.[21] In turn, almost 90 percent of Europeans say they recognize climate change as a major issue.[22]

Notably, popular magazines have joined respected science journals in reporting the problem. The cover of the May 2006 issue of *Vanity Fair* features both movie stars and politicians: here a rugged-looking Al Gore kneels near a wan Julia Roberts, who wears a crown of leaves and flowers; also featured on the cover are George Clooney and environmentalist Robert F. Kennedy. In May 2006, Davis Guggenheim's documentary about the dangers of climate change, *An Inconvenient Truth*, opened in Los Angeles and New York to critical acclaim and has gone on to win an academy award and become the third-highest grossing documentary in the United States to date.

In addition to the outcries of art and entertainment producers, Evangelical Christians, a major force in the conservative base, are now calling for a response to the crisis: forming an Evangelical Climate Initiative in February 2006, Christian leaders now are petitioning Congress to regulate greenhouse gases.

Investors are paying attention, too; *Fortune* journalist Marc Gunther warns: "Long-term investors, take heed: Global warming will have a significant impact on the financial performance of companies in your portfolio."[23]

Public awareness, public concern: this is why our team traveled to Antarctica. Just as the citizens of South Central Los Angeles worried about the impacts of the Safety-Kleen plant on their neighborhood, citizens of the world are now campaigning for corporate commitment to new, "green" systems of production—systems that take into account the changing climate of our earth.

Of course we care about the health of our planet. We also care, however, about how this public outcry will affect businesses in the United States. We care about how it will affect our American economy.

Politicians and business owners, if they are not directly affected by these climate changes, *are* affected by the will of the people. The day has come: industry must recognize the public's growing concern and under-standing about global climate change, or they may suffer the loss of public authority and profits. Responsible industry will help drive the debate toward real, global solutions that ensure total accountability.

Good leaders know that they need to be proactive to issues rather than reactive. And in the case of global climate change, industry is going to be forced to respond, if not now, then in the near future.

Let us begin with the public. What does the country know about global climate change? What aspects of the issue frighten or motivate them to call for action?

Today, global climate change is a problem that is taken seriously by many Americans. *TIME* Magazine / ABC News / Stanford University's 2006 poll seems to show this clearly. One summary of the reported information concludes that, of Americans polled:

> A large majority of Americans [. . .] (88%) think global warming threatens future generations. More than half (60%) say it threatens them a great deal; 38% feel that global warming is already a serious problem, and 47% feel that it will be in the future. Just over half of Americans (52%) say weather patterns in the county where they live have grown more unstable in the last three years and half (50%) feel that average temperatures have risen in their county. A large majority (70 %) think weather patterns globally have become more unstable in the last three years and 56 % feel average temperatures around the world have risen. Almost half (49%) say the issue of global warming is "extremely important" or "very important" to them personally, up from 31% in 1998. When asked about the causes of rise in the world's temperatures, 31% feel it is caused by the things people do, 19% feel it is due mostly to natural causes, and 49% feel it is a combination of the two. Almost seven-in-ten (68%) Americans think the government should do more to address global warming, according to the poll; however, 64% think scientists disagree with one another about global warming.[24]

Although many Americans believe that scientists disagree with one another on the issue of climate change, they do appear to believe that the phenomenon is occurring and that it is a problem that needs to be addressed. The numbers listed above are striking.

Knowing the position of the public can be an invaluable tool for business leaders, especially if the public sides with shareholders, and in light of media headlines having a lasting impact. If business owners want to have any role in shaping the solution to a public problem, then they need to become involved as soon as they discover that this problem is one for which the public will demand a resolution. As BP's chief executive, Lord Browne of Madingley, suggested, "the time to consider the policy dimensions of climate change is not when the link between greenhouse gases and climate change is conclusively proven, but when the possibility cannot be discounted and is taken seriously by the society of which we are part."[25]

Unfortunately, many businesses remain in a state of paralysis on the issue. Arguments made in the last two decades of the twentieth century convinced some CEOs either that the phenomenon was not occurring or else that they would not see evidence of climate change for decades (or even centuries). Because environmentalists and lawmakers spent so much time debating the forecasts put forth by climatologists, many listeners have dismissed climate change as a serious topic. Consider for example this excerpt from Dr. Patrick Jay Michaels's recently published book on climate change, *Meltdown: The Predictable Distortion of Global Warming by Scientists, Politicians, and the Media*. The author quotes Dr. John Christy in support of his theory that the climate isn't changing: "[The 2000 U.S. National Assessment of the Potential Consequences of Climate Variability and Change] is an evangelistic statement about a coming apocalypse [and] not a scientific statement about the evolution of a complicated system with significant uncertainty." [26]

Dr. Robert Thomas accompanied my team to Antarctica. While there, he attempted to explain to us the current stance of the academic world on climate change us. According to Dr. Thomas, "The scientific community has no doubt that it [climate change] is occurring. It is part of a natural process where we know if we go back in geologic time, we know that it is has been much, much warmer than it is today and we know that it has been much cooler than it is today. So we know that there are natural oscillations, but today, no doubt it is warming up." Dr. Thomas also points out: "There is some debate about what's causing that warm-up. We think the scientific community in general accepts that it's part of a natural response, a natural rhythm, and we think the scientific community accepts that man's use of greenhouse gases—especially of carbon dioxide—has exacerbated the problem."[27]

While the causes of the problem continue to be explored, over the past five years, more public debates on whether or not global climate change is actually happening have become increasingly rare. Many doubters have come to accept that our earth is experiencing a climate change. As *Time* journalist Jeffrey Kluger writes, "For years, popular skepticism about climatological science stood in the way of addressing the problem, but the naysayers [. . .] have become an increasingly marginalized breed."[28]

These days, it's not just scientists who heed the warnings. As Elizabeth Kolbert notes in an April 2005 issue of *The New Yorker*,

> In the same way that global warming has gradually ceased to be merely a theory, so, too, its impacts are no longer just hypothetical. Nearly every major glacier in the world is shrinking; those in Glacier National Park are retreating so quickly it has been estimated that they will vanish entirely by 2030. The oceans are becoming not just warmer but more acidic; the difference between day and nighttime temperatures is diminishing; animals are shifting their ranges pole-ward; and plants are blooming days, and in some cases weeks, earlier than they used to. These are the warning signs [. . .], and while in many parts of the globe they are still subtle enough to be overlooked, in others they can no longer be ignored.[29]

People are noticing changes when they travel and when they go out into their backyards. The weather seems a little bit different; the plants and the water seem a little bit different. Droughts, heat waves, and hurricanes seem a little more disastrous.

According to the world's second largest reinsurance company, Swiss Re, the cost of natural disasters exceeded $225 billion in 2005. This number was up from the previous record of $118 billion in 2004. One Swiss Re executive concluded, "Global warming has accelerated from a problem that might affect our grandchildren, to one that could significantly disturb the social and economic conditions of our lifetime."[30] Additionally, in a 2004 report on global climate change funded by Swiss Re and The Center for Health and the Global Environment at Harvard University, researchers claimed, "There is a danger that human intervention will accelerate and intensify natural climate changes to such a point that it will become impossible to adapt our socio-economic systems in time. [. . .] The human race can lead itself into this climatic catastrophe—or it can avert it."[31]

As climate change begins to have a visible effect on national economies, on companies, and on the environment, the public has begun to call for action.

What about American businesses? If we were to assess their needs, what would they be?

Most obviously, they will need to run themselves efficiently and profitably. Secondly, but also importantly, they will need to develop an ability to adapt to (and even thrive in) changing environments, be they global, economic, or social.

The reality of the global climate change debate is that the discourse has moved beyond whether this phenomena is real to whether and to what extent is the anthropogenic impact on the changes. In other words, what role does mankind play in the changes we are experiencing?

The public would like businesses to respond to the issue of climate change, and businesses can do this in such a way as to reap particular benefits. A number of American corporations feel that "going green" will allow them to run more efficiently, to garner public support, and to prepare for the future.

A number of American corporations have been planning for the consequences of global climate change for years now. A cover story for *BusinessWeek*, printed as long ago as August 2004, claims, "Remarkably, business is far ahead of Congress and the White House. Some CEOs are already calling for once-unthinkable steps. [. . .] Even without mandates, scores of companies are taking concrete actions."[32]

One company that has recognized the tremendous value and profound economic benefits of progressive environmental stewardship is Nucor Corporation. Nucor pioneered the electric arc furnace (EAF) technology, which allowed for the widespread and profitable recycling of used steel. The primary application of the EAF is for the melting of steel scrap; however, EAFs can be charged with limited amounts of iron scrap, pig iron, and direct reduced iron. EAF technology conserves more energy and is more protective of the environment than traditional steelmaking processes.

About 50 percent of total steel produced in the U.S. comes from EAFs in operations called minimills. Thus, the electric arc furnace has become indispensable to recycling such junk as old bicycles, worn-out refrigerators and crushed cars into useful products. At Nucor alone, the EAF process conserves 2,500 pounds of iron ore, 1,400 pounds of coal and 120 pounds of limestone for every ton of steel produced, annually saving enough energy to electrically power the city of Los Angeles for eight years.

Nucor's dedication to the environment, captured in its slogan "It's Our Nature," is one reason the American Steel and Iron Industry has so drastically outpaced emission reduction guidelines set by the Kyoto Protocol. The protocol called for carbon emissions to be 5 percent below 1990 levels by 2012. The American Steel and Iron Industry had reduced emissions by 36 percent of 1990 levels by 2003.

Nucor has also been extraordinarily successful in developing production methods that are efficient and environmentally friendly. Nucor recently announced a plan to produce pig iron in a way that will actually remove carbon dioxide from the atmosphere. The company is developing a pig iron facility in Brazil that will be fueled by charcoal produced from eucalyptus plantations near the plant. Planting and growing the trees that fuel the facility results in a net reduction of carbon dioxide from the atmosphere, an environmental breakthrough for pig iron production. The project will remove about 2,400 pounds of carbon dioxide from the atmosphere for every ton of pig iron that it produces, compared to the conventional method of production that increases the carbon dioxide in the atmosphere by 4,180 pounds for every ton produced.

Nucor's record of environmental stewardship is well known throughout the industry and in the business community at large. In 1991, then-CEO F. Kenneth Iverson was awarded the Medal of Technology by President George H.W. Bush for pioneering this process, which recycles a valuable resource and requires far fewer raw materials to be extracted from the environment for steelmaking. In 2005 the company was cited by Financial Executive magazine as being one of a handful of corporate leaders in highlighting social responsibility through its environmental practices. At a time when few mandates existed in promoting "green" technology, Nucor was spotlighted for its proactive stance.

In her 2006 address to the 17th Annual Energy Efficiency Forum, New York Sen. Hillary Rodham Clinton noted, "Last year in Auburn, New York, Nucor Steel spent $14 million on a continuous reheat furnace. Now that's money it will earn back in just three years through efficiency and higher productivity."

In 2002, the U.S. Environmental Protection Agency named Nucor to its Elite National Environmental Performance Track program. Then-EPA Administrator Christie Whitman said the program encourages businesses and public facilities to make voluntarily environmental improvements, above and beyond state and federal laws. The goal is to benefit the environment, people and communities while recognizing those who make the voluntary improvements. Nucor was accepted to the program based on its recent environmental success, the EPA said.

Nucor has found remarkable success meshing environmental stewardship with successful business practices. A July 2006 *Charlotte Business Journal* article cited the company's continued growth exceeding analysts' expectations and its dedication to forward thinking environmental projects. Nucor was named the top company in the BusinessWeek 50 in 2005 for exceeding other domestic steelmakers through a strategy of acquisition and growth coupled with increased productivity. The company was also named one of only 11 "Good to Great" companies in *Fast Company's* 2001 article. The magazine cited, among others, Nucor for "dramatic, remarkable, good-to-great corporate transformations."

In each of these transformations, "we found the same thing: There was no miracle moment," the article says. "Instead, a down-to-earth, pragmatic, committed-to-excellence process—a framework—kept each company, its leaders, and its people on track for the long haul. In each case, it was...the victory of steadfast discipline over the quick fix."

Flint Hills Resources, LP (FHR) is a wholly owned subsidiary of Koch Industries, Inc., and a company that has had great success integrating environmental responsibility into a tremendously successful business model. FHR supplies petroleum products to a large area of the Midwest from its Pine Bend refinery in Pine Bend, Minnesota, just outside of the Twin Cities. Pine Bend is among the largest and most technologically advanced refineries in the world.

From working to protect and conserve the delicate ecosystem of the Mississippi River, to partnering with the Science Museum of Minnesota to bring science into children's lives through exciting educational opportunities, FHR understands that investment in the community is an essential component to a long-term business strategy. This is particularly evident in FHR's significant investment in research and implementation of environmentally friendly technology.

Exemplifying its vision statement "Be Safe, Be Clean and Create Value," FHR has become a leader in producing cleaner burning fuels. FHR's Blue Planet® is a low-sulfur gasoline that cuts emissions by 10 to 20 percent from all vehicles and met the EPA's 2006 sulfur limit six years early. Blue Planet® has been recognized by the American Lung Association of Minnesota as a Clean Air Choice and listed as part of the Minnesota Pollution Control Agency's 10-point plan to cut air pollution. Blue Planet® has 70 percent lower sulfur content than the current national average and use of the gasoline is equivalent to removing 40,000 cars from the Twin Cities region.

FHR is currently working to extend this type of advanced and earth-friendly technology to reduce emissions from diesel fuel. What's more, FHR has made the investment into producing cleaner burning fuels while simultaneously reducing emissions at the Pine Bend facility by 50 percent over five years. FHR is managing to prove that in today's economy, environmental responsibility goes hand-in-hand with long-term business planning.

Truth be told, many of these companies have been pleased to find that adopting "green" technology and strategies can be both sensible and profitable.

One benefit is a new freedom from the Persian Gulf and other sources of imported oil. If American companies can design cars and factories that are more energy efficient or that that utilize alternate sources of energy, both the companies and the country can begin to limit our dependence on oil. A focus on polluting less can also lead to a renewed focus on efficiency and to the development of innovative new techniques and/or technology. According to Michael Northrop, co-creator of the Climate Group, a coalition of businesses and governments dedicated to cutting greenhouse gases, "It's impossible to find a company that has acted and has not found benefits." Jim Walker, the Climate Group's chief operating officer, claims that at least 74 major corporations have reduced their greenhouse gas emission levels since 1990 and that most of these corporations realized a profit from the reductions; in turn, 17 U.S. cities saved $600 million in 2004 from greenhouse gas emission measures.[33] "Early movers will gain the long-term advantage," said Walker.[34]

Because Washington has been slow to act on the issues, states and businesses have been leading the way. "States have stepped up to fill this policy void, as much out of economic self-interest as fear of devastating climate changes," said Kenneth A. Colburn, executive director of Northeast States for Coordinated Air Use Management.[35] In March 2005 the Northeastern states of Connecticut, Delaware, Maine, Massachusetts, New Hampshire, New Jersey, New York, Rhode Island and Vermont took up a "Regional Greenhouse Gas Initiative" in an effort to establish limits on carbon dioxide emissions and to allow companies that work below standard limitations to sell credits to other firms. In addition to lowering overall compliance costs, this initiative was designed to help businesses function more efficiently and economically.[36] Correspondingly, the West Coast Governors´ Global Warming Initiative, launched by the governors of Washington, Oregon, and California in September 2003, is considered to be leading state initiatives on climate change. The Governors of these West Coast states "have committed to act individually and regionally to reduce greenhouse gas emissions below current levels through strategies that promote long-term economic growth, protect public health and the environment, consider social equity, and expand public awareness."[37] Additionally, on June 13, 2005, the mayors of more than 200 cities signed the U.S. Mayors Climate Protection Agreement, pledging to meet the Kyoto goal of reducing greenhouse gas emissions in their cities to 1990 levels by 2012. But we need to be careful to demonstrate real leadership, thoughtful and meaningful, not leading with our chin. Our leadership must involve a global response, not a fragmented local response, which could drive industry into areas of unregulated environmental conditions.

Businesses, too, are looking to the future. In 2004, DuPont saved hundreds of millions of dollars by reducing its greenhouse gas emissions to 72 percent below 1990 levels (significantly exceeding its goal of a 65 percent reduction by 2010). In turn, DuPont developed new green products; specifically, the Tyvek® sealed roof system saves energy and "therefore reduces the associated CO_2 emissions by 11 tons over the lifetime of the building."[38] The Whirlpool Corporation also has committed to decrease the company's greenhouse gas emissions (from global manufacturing, product use, and end-of-life waste). As CEO Jeff M. Fettig claims: "At Whirlpool Corporation we take our environmental responsibilities very seriously. Just as we have taken a global approach to our home appliance business, we believe our world's environmental issues, such as climate change, must be addressed in a similarly comprehensive way."[39]

In 2005, GE Chief Executive Officer Jeffrey Immelt called for the United States to set clear goals on carbon emissions. Immelt claims: "it's no longer a zero-sum game—things that are good for the environment are also good for business." Immelt tells shareholders that GE has determined to take on this initiative "not because it is trendy or moral, but because it will accelerate [economic] growth."[40] In addition to reducing companywide greenhouse gas emissions and by improving energy efficiency, GE also plans to profit from selling wind turbines and more efficient locomotives and jet engines. Additionally, GE plans to introduce new, safer nuclear power plants.

GE is not the only power company that has begun to make changes. *Businessweek* recounts the experiences of American Electric Power:

> American Electric Power Co. (AEP) once fought the idea of combating climate change. But in the late 1990s, then-CEO E. Linn Draper Jr. pushed for a strategy shift at the No. 1 coal-burning utility— preparing for limits instead of denying that global warming existed. It was a tough sell to management. Limits on carbon emissions threaten the whole idea of burning coal. But Draper prevailed. Why? "We felt it was inevitable that we were going to live in a carbon-constrained world," says Dale E. Heydlauff, AEP's senior vice-president for environmental affairs. Now, AEP is trying to accumulate credits for cutting CO_2. It's investing in renewable energy projects in Chile, retrofitting school buildings in Bulgaria for greater efficiency, and exploring ways to burn coal more cleanly.[41]

This type of story is playing out again and again among many corporations that are looking to the future. Wal-Mart now has plans to install wind turbines on some of its stores to generate electricity. The world's second largest bank, HSBC, recently pledged to neutralize its carbon output by investing in wind farms.

Consider further the role and opportunity of renewable energy. In particular, how harvesting the invested energy in scrap from renewable products like steel, paper, and glass can help reduce the burden of landfilling, while helping reduce greenhouse gas emissions.

The nation's energy challenges -- the need for energy independence and security, environmental concerns with depleting our natural resources, and recent electricity supply shortages have necessitated the shift toward renewable energy development and legislative mandates such as renewable portfolio standards ("RPS").

A RPS requires certain electricity sellers and/or buyers to maintain a prescribed percentage of designated clean resources in their wholesale supply mix.[42] Portfolio standards are flexible in that certain technologies can be included in the renewables definition, or certain subgroups of technologies can be targeted for inclusion at distinct levels.[43] The standard allows market competition to decide how best to achieve these standards. The standards become self-enforcing as a condition of retail sale licensure.[44]

The advantages of a RPS is that it does not subsidize any particular technology or locus of that technology, there is no government subsidy of any technology or project, the market dictates which renewable technologies and projects are actually successful based on competition, and non-compliant owners can purchase credits from others rather than construct renewable projects themselves.[45] RPS requirements can be applied under any wholesale or retail competition, without placing any entities at a disadvantage.[46]

The role of renewable energy programs in increasing energy security gained new prominence as oil and gas prices rose dramatically late in the year 2000 and again in 2006.[47] The terrorist attacks of September 11, 2001, the Blackout of 2003, the most recent war in Iraq, and Hurricanes Katrina, Rita, and Wilma in 2005 led to heightened concerns about energy security, the vulnerability of energy infrastructure, and the need for alternative fuels.[48] Further, the 2001 electricity shortages in California brought a new emphasis to the role that renewable energy may play in electricity supply.[49] Worldwide focus on the environmental problems caused by air and water pollution and global climate change, and the related development of clean energy technology in western Europe and Japan, have also heightened awareness of the necessity for renewable energy sources.[50]

Renewable energy can help address multiple problems, including volatile fossil fuel prices, energy supply shortages and disruptions, a growing dependence on natural gas, a need for more domestic energy supplies, and harmful air pollution. It will help stabilize electricity prices, reduce natural gas prices, reduce emissions of carbon dioxide and other harmful air pollutants, and create jobs, especially in rural areas.[51]

United States Congressman Joe Barton addressed this issue when he called for recycling to be included in the nation's renewable energy policy:

> Mr. Speaker, I rise today to call attention to the energy security issues we face and to highlight the important role renewable energy plays in producing reliable electricity and curbing demand for power transmission equipment. Given this importance, we must continue to support initiatives to improve transmission, increase generation nationally and encourage renewable energy and conservation.
>
> Effective renewable energy policy must include the significant invested energy available through greater recycling. Manufacturing recycled products requires, on average, 17 times less energy than manufacturing the same products from virgin materials. In addition to the traditionally understood benefits of recycling as a conservation and waste management tool, recycling is becoming increasingly understood as an energy source available to combat the Nation's growing energy crisis. [. . .] [R]ecycling should be an integral component of our Nation's energy efficiency strategy.
>
> The Federal Energy Policy Act of 2005 acknowledged the high invested energy content of recyclables. Section 1353 mandated the U.S. Secretary of the

Treasury, in consultation with the Secretary of Energy, to conduct a study to determine and quantify the energy savings achieved through the recycling of glass, paper, plastic, steel, aluminum, and electronic devices, and to identify tax incentives that would encourage recycling of such materials. As chairman of the House Energy and Commerce Committee, I was deeply involved in the creation of this legislation, which I am proud to say was achieved with strong bipartisan participation and support.

The Environmental Protection Agency has shown that recycling saves billions in energy costs and hundreds of millions of tons of raw materials. American companies that engage in the use of renewable resources and recyclable materials are already contributing countless billions of dollars per year to our Nation's energy grid. These companies are engaging in practices that are environmentally friendly and energy conservative.

By harvesting the invested energy in recycling, this country saves tens of millions of tons of ore, coal, trees and billions of dollars in energy costs. Recycling programs offer a means to conserve natural resources, ease the burden on the grid, reduce excessive municipal waste, protect the environment, create jobs, and save energy.

Like the energy drawn from wind or from water behind a dam, recyclable materials contain a vast amount of energy that is available to be harvested. Recyclables possess invested energy, and in order to remain competitive, we must focus on capturing the energy that is already invested in recycling. Through recognition of these policy objectives, we will create the most efficient market-based solutions to ensure a safe, abundant, and stable energy supply to our citizens for years to come.[52]

While the disposal practices of landfilling and incineration have improved significantly over the last three decades, there are still substantial environmental risks associated with these activities. Recycling is a critical policy priority for the reduction of solid metals and other substances which contribute to the amount of leachate in landfills.[53] In addition to leachate, methane and other volatile greenhouse gasses are generated inside landfills.[54] Capture and reuse of this methane has provided a valuable source of renewable energy.[55] Landfills are a low-priority form of waste management[56] and a poor use of land, which might be used in other more productive ways.

By reducing the amount of energy used by industry, recycling reduces greenhouse gas emissions and helps prevent global climate change.[57] This is because much of the energy used in industrial processes and in transportation involves burning fossil fuels such as gasoline, diesel and coal, the most important sources of carbon and other greenhouse gas emissions.[58] Additional benefits are derived from reduced emissions from incinerators and landfills and by slowing the harvest of trees, which function as carbon sinks.[59]

In addition to greenhouse gases, recycling can reduce a range of pollutants from entering the air and water.[60] By decreasing the need to extract and process new raw materials from the earth, recycling can eliminate the pollution associated with the initial stages of a product's development: material extraction, refining and processing.[61] These activities pollute the air, land and water with toxic materials, such as ammonia, carbon monoxide, methane, and sulfur dioxides.[62] Further reductions are achieved as a result of energy saving, thus reducing greenhouse gas emissions and other air pollutants.[63]

By converting waste into valuable products, recycling creates jobs, contributes feedstock to manufacturing, and adds significant value to the economy. Nationally, nearly 1.3 million jobs have been created in more than 56,000 public and private facilities.[64]

Recycled paper and paperboard mills employ 139,375 people, and recycling steel mills employ 118,544 people.[65] These are manufacturing sector jobs and essential to state and local economies reeling from the effects of outsourcing. Furthermore, the average annual wage per recycling sector job is nearly $36,000,[66] close to the top tier of all manufacturing jobs[67] and significantly higher than the national average salary.[68] In total, recycling facilities add $236 billion in gross annual sales and produce more than $37 billion in annual payroll.[69] These numbers are based on a relatively low national recycling rate of 28 percent of solid waste.[70]

In addition to the direct benefits, support businesses that provide goods and services to the recycling and reuse establishments also contribute to the economy. These supporting activities include recycling and reuse equipment manufacturers, consulting and engineering services, brokers, and transporters.

Policies which realize the high energy content of recyclables have long been acknowledged by federal agencies and through federal legislation. In January 1974 the EPA published a position paper in response to the energy crisis precipitated by the Arab oil embargo of 1973. In the paper, the EPA stressed the need for resource recovery programs by stating, ". . .for many materials the energy required to mine the ore, refine it, and produce a finished product is much larger than that needed to reclaim the materials from wastes. . . .The potential savings from recycling steel, aluminum and paper is equivalent to about 190,000 barrels a day."[71]

Thirty-one years later, FEPA similarly acknowledged the high invested energy content of certain recyclables. Section 1353 mandated the U.S. Secretary of the Treasury, in consultation with the Secretary of Energy, to conduct a study to determine and quantify the energy savings achieved through the recycling of glass, paper, plastic, steel, aluminum, and electronic devices, and to identify tax incentives that would encourage recycling of such materials.

The amount of energy wasted from failure to recycle aluminum and steel cans, paper, printed materials, glass, and plastic translates to the annual output of 15 medium sized power plants. "The reuse of 500 steel drums per week yields 6 trillion Btu per year, which is enough energy savings to power a city the size of Colorado Springs, Colorado for one year. . . As a result, recycling should be an integral component of our nation's energy efficiency strategy."[72]

These materials—steel, aluminum, paper and glass—are essential ingredients in our industrial economy.[73] Steel is perhaps the most foundational material to our industrial life. Steel is to our infrastructure and general manufacturing as oil is to our transportation.[74] Our choice is either to produce such materials domestically or to import them. In the case of scrap steel, if we don't recycle scrap into new steel we are wasting energy, if we export it, we are actually exporting a source of energy.

As global climate change begins to have more visible effects, other industries will be compelled to change, even if they have not chosen to address the problem ahead of time. Even if the global community manages to make significant cuts in emissions, climatologists claim that the earth will continue to warm several more degrees in the approaching decades. Additionally, one thing that scientists have discovered as they have explored the phenomenon of global climate change is how swiftly climate changes can take effect. For many years, the thought was that our population would have centuries to address this problem. Unfortunately, experts now know that this is not the case.

The results of global climate change will affect agricultural production, raise sea levels, and bring more extreme weather. Consequently, insurers will have to cope with more floods, storms, and other disasters. Growers will modify their farming techniques to respond to changing water levels and temperatures. As *BusinessWeek* notes, "Companies that pioneer low-emission cars, clean coal-burning technology, and hardier crop plants—or find cheap ways to slash emissions—will take over from those that can't move as fast."[75]

What will the effects be on the average businessman, the average citizen? Eugene Linden of the *Los Angeles Times* writes:

> [T]he real, more insidious scenario might be that climate change will intrude on our lives like an omnipresent and ever more confiscatory taxman. Where they can, insurers and banks will pass weather risks to individuals and the government, making the costs of daily life more expensive. In some areas, housing might become uninsurable and unsalable, which in turn could cause a financial crisis. Municipal budgets and government safety nets will gradually succumb to the ever-increasing burden imposed by windstorms, floods, droughts and other weather extremes. Infectious diseases will thrive. The middle class will slowly find its savings and creature comforts stripped away, and the ordinary details of living, such as eating fresh vegetables and traveling to see family and friends, will become more expensive and uncertain. At some point it will dawn on us that the weather is making us poorer and sicker.[76]

The possibilities are many.

Although some CEOs have remained hesitant, assuming that they can adapt to the changes as they come, other business leaders have realized that the risks of this type of "strategy" far outweigh any present benefits that they may provide. It makes little sense to run the risk of being unprepared for catastrophic changes when valuable precautions can be taken now, and at a small price. As the 2004 issue of *BusinessWeek* claimed, "A British government panel, for instance, concluded that the cost of its share of the task of limiting the level of CO_2 to 550 ppm would be about 1 percent of Britain's gross domestic product. Compare that, says Sir David King, with the cost of a single flood breaking through the barrier in the Thames River—some £30 billion, or 2 percent of current GDP."[77] It is advantageous to recognize future problems and to plan for them accordingly. As economist Paul R. Portney, president of Resources for the Future, claims: "Common sense says that it's time to purchase some low-cost insurance now."[78]

If corporate America doesn't step up, business is likely to suffer. If businesses do step up, they are likely to garner public applause and to reap financial benefits. However, the situation for American corporations remains complex. This is because of the world's failure to take a sound stance on the issue.

American industry currently is suffering as a result of an uneven playing field. The Kyoto Protocol has been the most highly publicized attempt by the global community to address the issue of climate change. However, under Kyoto, only developed nations (such as the United States and the countries of Europe) are asked to reduce their emissions (particularly carbon dioxide) to lower target levels. If they cannot do so, they must buy emission credits or invest in conservation. But critics of Kyoto assert that this is in fact not a true "global solution." Developing economic powerhouses, like China and India, have no such immediate restrictions under Kyoto.

This is one issue that truly concerns our team. Simply put, we believe that the United States cannot resolve this issue on its own, and we don't feel that American businesses should be punished for going green because polluters in other countries can produce cheaper goods out of their dirty factories.

This problem is a global problem. Any *partial* solution is a *failed* solution.

Insisting on stricter standards for developed nations only means that corporations are moving jobs to places where there are no standards. This is harmful. Harmful not only to the environment, but also to the American workforce and to the American economy.

Nucor Steel's General Manager for Environmental Compliance, Steve Rowlan, explains: "In my opinion, they've really got Kyoto backwards. If you were to reduce production in the developed nations and shift it to a non-developed nation, the amount of carbon emissions would actually go up and result in a loss to the environment. Also, developed nations would lose jobs associated with their contribution to Kyoto."

But how can we devise a global solution to this global problem? As GE's CEO Jeffrey Immelt recently claimed, his company's voluntary effort is by no means a substitute for government action. In May of 2005, Immelt petitioned federal leaders to begin by "clarifying policy" and "committing to market mechanisms," claiming, "Industry cannot solve the problems of the world alone—we need to work in concert with government."[79]

How can industry move our government to demand stricter global environmental standards? Currently, it seems that our government, as well as those governments of other progressive nations, aren't willing to do what it takes to promote change. So far, the United States has refused to change its reliance on a place like China in order to take a leadership role—a role that would benefit both the corporations of America and the people of the world.

The key lies in educating the public. The public needs to know, not just that global climate change is a real and significant problem, but that current international proposed solutions aren't good for American businesses. Once educated, the public can speak out on the issue of climate change: they can call and write their representatives, they can ask questions of lawmakers, they can voice their opinions through their vote. If they are given the tools to speak knowledgeably on the issue, they can in turn call for a global solution.

I think back to Juanita Tate and the way she invigorated the community of South Central Los Angeles. I look at current grassroots political movements and see the way that they inspire the public to change their votes—and to change the political makeup of this country. If businesses want change, they need to get people to want change, too. And you know what? This is easy to do when people and businesses want similar things.

Once businesses and people work together, they can bring about many mighty changes in this world. As Albert Einstein once wrote, "There is no scientific antidote . . . only education. You've got to change the way people think. [. . .] What I want to do is to disarm the mind. After that, everything else will automatically follow. The ultimate weapon [. . .] is international education."[80]

How does one inspire public awareness? One goes to Antarctica to see the problem head on. One speaks to the people—in town hall meetings, on public television, in their homes and workplaces. I say you don't do politics over the telephone. You don't deal with environmental remediation over the telephone or read about it in the books. You get down, roll your sleeves up first hand, and take a look—that's what this process is all about.

As lawyers and political scientists, we are social scientists, not natural scientists. We are not involved in challenging the precise "hard" science behind global climate change. We are, however, concerned with gathering reputable data, with charting and organizing the various arguments and positions being taken on climate change today, and with measuring the credibility these theories are given by the public. We want to assist companies and our communities as they confront the demonstrated views of the broader public. With this issue, it is overwhelmingly clear that the public believes that there is reason for valid concern. For those whose businesses intersect with government—the instrument of the people—they either will engage with these issues or else they will be forced to deal with the forthcoming policies that will result from this palpable public outcry.

So, in the next chapter, I want to equip you with the some of tools you'll need as you consider why and how American businesses need to rethink the way we (as a global community) are addressing the problem of climate change. I feel that my experience with businesses and communities can help to form plans of action that will allow corporations and their shareholders to benefit, while at the same time allowing them to provide excellent community environmental stewardship.

Chapter 4

What is global climate change? What are its effects?

Climate change is nothing new—the Earth's climate has varied dramatically in the past and it will continue to do so in our future. However, past changes have generally occurred slowly over time and in response to various geologic factors. The recent concern is that that manmade greenhouse gasses (carbon dioxide, methane, and others) are influencing and exaggerating the natural cycle of climate change. Greenhouse gasses, especially carbon dioxide, act like a blanket that wraps around our earth. When the volume of greenhouse gas increases (due to natural environmental cycles as well as the burning of fossil fuels like coal, oil, and natural gas), the blanket gets denser and the planet heats up.

And as the planet heats up . . . the polar ice caps begin to melt. Sea levels rise, and coastlines change. More inland tracts of land are devastated by drought and wildfires. Tropical storms worsen.

Populations suffer.

For a long time, global climate change wasn't a popular issue, perhaps because scientists always seemed to be arguing about it. However, as Dr. Robert Thomas explains: "The scientists' argument is not whether it's happening or not or whether mankind has something to do with it. Their argument is about what speed is it happening, what direction is it taking, what are the trends, what's the future based on what's happening now."[81]

Generally, the relative warmth or coolness of the planet is influenced by the composition of our atmosphere (and the relative amounts of water vapor, carbon dioxide, methane, sulfur dioxide, and various other gases and particulates that compose it), by changes in the Earth's orbit around the Sun (and possibly by changes in the Sun's orbit around the galaxy), by the motion of the planet's tectonic plates, by variations in the Sun's solar output, by the orbital dynamics of the Earth-Moon system, by the collision of the Earth with significantly sized meteorites, and by the eruptions of supervolcanoes.

Most of these forces exist beyond the control of man. Nevertheless, we are able to pollute our air. And, by putting large quantities of greenhouse gases into the atmosphere, we put new pressures on our climate system. These pressures may lead to drastic environmental changes—

changes that are able to take place without much prior warning. As the geologist W.S. Broecker once noted, "Climate is an angry beast, and we are poking it with sticks."[82]

Human contributions of greenhouse gases come largely from three sources: industry, power, and other sources—largely transportation and residential uses.

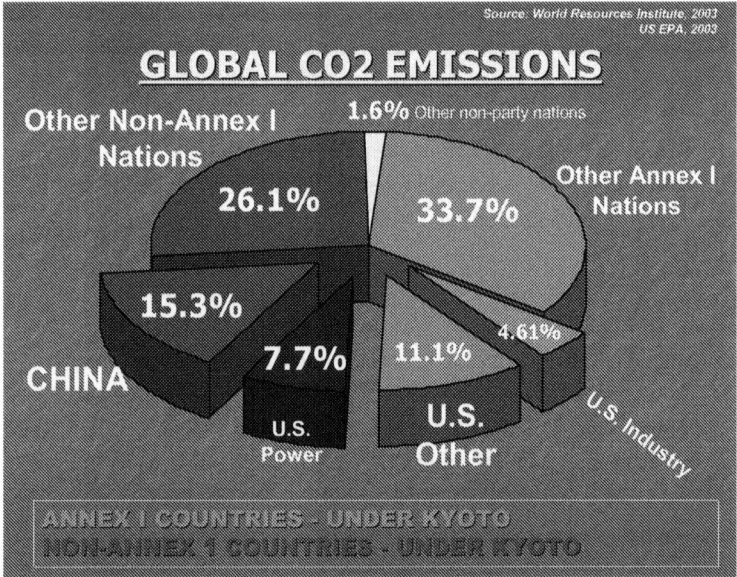

Fig. 4-1: Total Global CO-2 Emissions

It is interesting, however, that most debate centers on regulating emissions from industry or power utilities, and fail to address the largest category affecting the environment—transportation and residential emissions. One might opine that this is because the impact might be squarely on the back of us—the voters. However, within this conundrum lies opportunity. It is within this category of emissions that we could see dramatic improvement through developments in such technology as residential solar power usage and other alternative fuels for our homes and transportation. These improvements would not only reduce emissions from the largest sources in the United States, but would also fuel economic development as consumers invest in these new technologies.

Let me take a second and address what exactly is occurring.

Carbon dioxide is an important component of our atmosphere. Indeed, it's what keeps Earth warm enough for human survival and for plants and animals to flourish in the different climates produced by our planet. But too much carbon dioxide can have disastrous effects.

Carbon dioxide is a powerful compound. The gas represents just a few hundred parts per million (ppm) in the overall air blanket that surrounds our Earth, but it is carbon dioxide that that a) allows sunlight to stream in and b) stops heat from radiating back out.

Too little carbon dioxide results in an ice age. By analyzing the carbon dioxide trapped inside Antarctic ice core samples, scientists have discovered that 19,000-17,000 years ago, the atmosphere's CO_2 concentration was just 180 ppm. During that time, large sections of our planet were covered by ice.

After the glaciers retreated, the carbon dioxide level rose to 280 ppm. However, since the start of the Industrial Revolution, carbon dioxide levels have jumped from 280 ppm to about 360 ppm today.

This jump has had notable consequences. As *TIME* journalist Jeffrey Kluger writes, "Of the 20 hottest years on record, 19 occurred in the 1980s or later. According to NASA scientists, 2005 was one of the hottest years in more than a century."[83] We can see the consequences on ice, land, through fires, and in our waters.

ICE

In Antarctica, one can see quantifiable, demonstrable evidence that the earth's surface is warming. That's why I traveled there with my team. The Antarctic Peninsula, and specifically the west coast of the peninsula, is warming two to three times faster than the global average, and the average annual temperature of this region has increased almost 5 degrees Fahrenheit in the past 50 years.

Fig. 4-2 Larsen B ice shelf, 17th Feb 2002

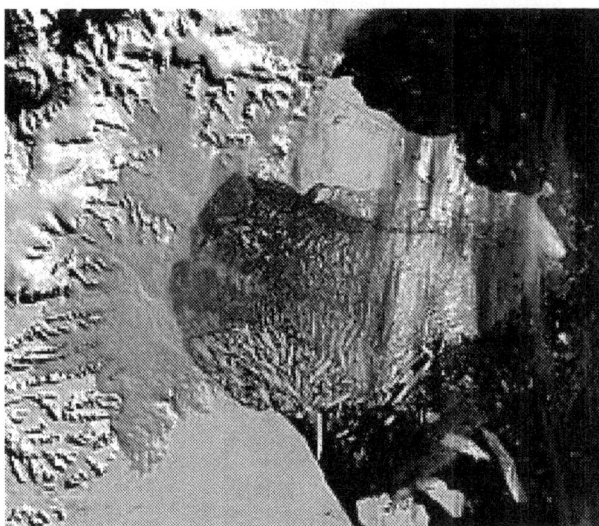

Fig. 4-3: Larsen B ice shelf, 5th March 2002
(16 days later)

Courtesy of NASA and Visible Earth (http://visibleearth.nasa.gov/)

One thing we saw on our trip was the way in which the Larsen B ice shelf has visibly deteriorated. Once about 220m thick (720 feet), the shelf lost about 3,250 km² of ice into the ocean during a 30 day period in the Spring of 2002. Prior to its collapse, the massive shelf was thought to have stood on the coast of Antarctica for the past 400 years and perhaps even as long as the past 12,000 years (since the end of the last ice age). Today it exists only as a chaotic, turbulent sea of churning water and deadly, projectile islands of ice.

The speed of the shelf's disintegration is shocking. In fact, the current remains of the Larsen B shelf represent about 40 percent of the body of ice that existed there back in 1995.

How did global climate change bring about the break up of the ice shelf?

Higher temperatures, as well as large amounts of summer melt-water, caused the ice to break up and liquefy. In particular, scientists have observed that the melt-water had a tendency to run down the crevasses in the ice shelf. The movement and pressure of the melt-water on the ice caused the shelf disintegrate even more quickly.

Frightening but true: The Larsen B ice shelf is not the only part of Antarctica that's breaking up. Overall, seven ice shelves in the Antarctic Peninsula have declined in area by about 13,500 km² since 1974.

What are the results of this ice breakup?

For one thing, the decline in Antarctica's krill population appears to be related to the decline in the amount of sea ice on the continent. Krill (*Euphausia superba*) are tiny, shrimp-like creatures eaten by whales, seals, birds and fish. Studies conducted in November 2004 illustrated how populations of krill in Antarctica have dropped dramatically in recent years. Krill numbers may have gone down by as much as 80 percent since the 1970s, causing today's krill population to subsist at 1/5 the size that it was 30 years ago. Not surprisingly, the decline in krill affects the other species that live in Antarctica.

Smaller populations of krill may account for the smaller populations of some penguin species. Specifically, environmentalists have growing concern about Antarctica's Adélie penguins, which eat mostly krill and which currently are abandoning former nesting grounds and shrinking in number.

The decline in krill has also made it more difficult for the great baleen whales to return to pre-exploitation numbers. This whale population is still suffering as a result of the hunting/whaling practices that took place between 1925 and 1975, and the decline in this particular food source only makes it more difficult for them to breed and survive.

When we traveled to Antarctica, we could tell that the cost of massive ice disintegration would be significant. When there is a decline in the population of krill, when there is a decline in the populations of larger species that feed on krill, the planet suffers, and so does business. The southern ocean is a valuable resource for fisheries, and many of the species of fish that are caught there subsist on krill. The fishing industry is not the only trade that utilizes the land and waters around the South Pole; thousands of tourists travel to Antarctica every year in order to enjoy the magnificent wildlife. That wildlife depends on krill for food and on a consistent landmass and shoreline for a habitat.

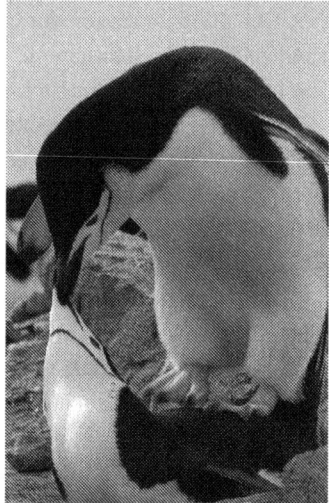

But let's stop and think about the bigger picture here for a minute. The cost of melting Antarctic ice is not limited to Antarctica. As that ice melts, it floods oceans with freshwater. The consequences? Rising oceans.

One of the most comprehensive current studies is the Arctic Climate Impact Assessment (ACIA), which warns that even small increases in sea level will have significant effects.[84] Such a rise "could be extremely serious, undermining coastal ecosystems, communities, and livelihoods."[85]

And it's not just the Antarctic ice shelf that is feeling the effects of global climate change. The North Pole and Greenland also offer significant illustrations of how quickly ice sheets can melt. BBC reporter David Shukman describes his trip to Southern Greenland's Sermilik Glacier:

First you hear a savage cracking sound, next the rolling crash of thunder. Then as the icebergs rip away from the margin of the ice-sheet they plunge into the grey waters of the Atlantic with a roar that echoes around the mountains. Nothing prepares you for the sheer scale and drama of events in this forbidding terrain and all the signs are that the changes at work here are gathering pace.[86]

Measurements of the Greenland ice sheet taken from passive microwave satellite sensors in June of 2006 show 685,000 square kilometers of melt. Indeed, the summer of 2006 was part of a record-breaking year of northern polar ice loss in Greenland. Carl Boggild and his team of NASA scientists have recorded falls in the height of the Greenland ice sheet at 10 meters a year; indeed, now in some places the ice is dropping at a rate of one meter a month. The melting of the Greenland glacier and surrounding sea ice, as well as the island's disappearing permafrost and northern expansion of vegetation, illustrate how dramatically a piece of land can respond to changes in atmospheric temperature. In turn, the movement of water over and into the glacier (as the ice melts) results in a proliferation of earthquakes—up to magnitude 5.0. In 2005, Harvard and Columbia researchers detected as many ice quakes as the total detected in the time between 1993 and 1996.[87]

And once the melting begins, it only accelerates. Excess water weight pushes down on glaciers at the same time that water seeps through cracks to the underside, promoting further meltdown.

LAND

The most frightening issue? Scientists do not know how the world may be affected by the combined melting of Earth's two major ice sheets, in Greenland and in Antarctica. These ice sheets hold 77 percent of the world's fresh water. This is enough water to potentially raise the sea level approximately 225 feet (70 meters). Not only would this influx of fresh water disturb the Gulf Stream; sea levels could rise dramatically. By some estimates, if the entire Greenland ice sheet were to melt, it could raise global sea levels 23 feet, swamping large areas of coastal Florida and most of Bangladesh. The Antarctic ice sheet contains enough ice to raise sea levels more than 215 feet.

Through computer models, scientists have demonstrated how different land masses would respond to a rise in sea level. In Florida, if melting ice caused a 6-meter (20-foot) rise in sea level, the oceans would swamp Miami, Fort Lauderdale, and Tampa. The entire Florida coastline, as well as parts of Orlando and other inland areas, would be severely affected.[88] According to James Gustave Speth, author of *Red Sky at Morning: America and the Crisis of the Global Environment*, "In the United States, Florida and Louisiana are especially vulnerable. In Bangladesh, seventeen million people live fewer than three feet above sea level. Small island nations in the Pacific and Indian Oceans are threatened, as is The Netherlands, where half the population lives below sea level."[89]

Scientists are using computer models to predict geographic changes that may potentially result from global climate change. But are businesses and lawmakers designing policy models to deal with these potential effects?

If massive changes in the shape of the Earth's continent seem like extreme future possibilities, one might instead look at some of the smaller but disturbing effects of global climate change that we can already see around us in our world today.

Since 1970, our oceans have warmed a full degree Fahrenheit.

Warmer water strengthens typhoons and hurricanes.

In the past 35 years, the number of Category 4 and 5 hurricanes worldwide has doubled; in turn, the wind speed and duration of all hurricanes has increased by fifty percent. 2005's Hurricane Katrina is certainly one example of the devastating effects such powerful storms can have on the people, land, and businesses of a thriving coastal city.

Climate models also indicate that severe inland droughts have become more common as temperatures have warmed. According to the National Resources Defense Council, "Greater evaporation, particularly during summer and fall, could exacerbate drought conditions and increase the risk of wildfires."[90] Higher temperatures draw moisture out of soil, causing dry regions to become drought regions. As *TIME* journalist Jeffrey Kluger explains, "People, animals and plants living in dry, mountainous regions like the western U.S. make it through summer thanks to snowpack that collects on peaks all winter and slowly melts off in warm months. Lately the early arrival of spring and the unusually blistering summers have caused the snowpack to melt too early, so that by the time it's needed, it's largely gone."[91] In turn, El Niño cycles, driven by warm Pacific waters, draw rain away from the dry areas of Africa and East Asia.

Drought destroys plants and harms animal populations. The drought brought on by global climate change is likely to destroy all kinds of animal species, from fish to amphibians to mammals:

> Environmental groups can tick off scores of species that have been determined to be at risk as a result of global warming. Last year, researchers in Costa Rica announced that two-thirds of 110 species of colorful harlequin frogs have vanished in the past 30 years, with the severity of each season's die-off following in lockstep with the severity of that year's warming. In Alaska, salmon populations are at risk as melting permafrost pours mud into rivers, burying the gravel the fish need for spawning. Small animals such as bushy-tailed wood rats, alpine chipmunks and pinon mice are being chased upslope by rising temperatures, following the path of the fleeing trees. And with sea ice vanishing, polar bears—prodigious swimmers but not inexhaustible ones—are starting to turn up drowned.[92]

Besides being devastating in principle, the loss of specific animal species can affect businesses and communities that depend on those species for profit or for a sense of communal identity. For example, in recent memory, the polar bear has become the symbol of Coca-Cola's Christmas advertising. The bear's environment has become so precarious that President Bush's administration has proposed listing Polar Bears as threatened under the Endangered Species Act.[93]

FIRE

Drought conditions, combined with the temperature increase, also affect wildfires. The longer, fiercer fires that have taken hold of lands in the western United States over the past two decades seem to be connected to a) warmer summer temperatures, b) earlier spring arrivals, and c) earlier snowmelts in the West. In 2006, climatologists analyzed a comprehensive U.S. government database of forest fires (all of which consumed more than 1,000 acres).[94] Looking at fires that took place between 1970 and the present, they discovered that after 1986 American fires increased in their intensity. After 1986, spring and summer temperatures were higher (by about 1.6 degrees Fahrenheit); consequently, the average snowmelt in the West occurred about a week earlier.

Earlier snow melts allowed the natural vegetation (and the soil) to dry out more quickly, and this allowed fires to begin earlier in the season.

Large fires, many of them burning through 6 1/2 times more area, were four times more frequent than during preceding years. Additionally, it was discovered that the average wildfire season increased by 2 1/2 months. As *Boston Globe* reporter Naila Moreira writes, "The new findings suggest that the most up-to-date forest management methods may be insufficient to slow the uptick in large forest fires. Most climate researchers believe that global temperatures will continue to warm as human activity increases the concentration of heat-trapping greenhouse gases, such as carbon dioxide, in the atmosphere."[95]

Before this study, scientists believed that the increased force and duration of Western wildfires were caused by specific kinds of land use. Specifically, it was thought that the general suppression of forest fires had, over a period of years, caused dead and dry vegetation to accumulate in Western forests, making more fuel available for the fires. However, this recent study illustrates how a large percentage of particularly bad wildfires have burned not in the West but in the Northern Rocky Mountains—an area where there have been few land-use changes. Scientists concluded that it's the change in the climate—not the change in land use—that's resulting in hotter, faster fires. Indeed, scientists discovered that "66 percent of the yearly variation in forest fires could be explained by temperature changes alone, with hotter years producing more fires."[96]

The increase in wildfires is caused in large part by the earlier snowmelts; in turn, the earlier snowmelts are caused by the change in the Earth's climate.

WATER

This text generally focuses on changes in atmospheric temperature, a condition that produces melting glaciers and stronger storms, as well as a rise in the sea level. However, I also want to draw attention to some other deleterious effects that unregulated industry has on our planet. Most notably, I would like to point to the 2006 study sponsored by the National Center for Atmospheric Research and the U.S. Geological Service.[97] The multi-agency report was compiled by scientists from Australia, Canada, France, Germany, Japan, Monaco, New Caldonia, and the United States. Entitled *Impacts of Ocean Acidification on Coral Reefs and Other Marine Calcifiers*, the report points out that we are now seeing an acidification of our Earth's oceans. According to the report, the world's oceans have absorbed approximately 118 billion metric tons of carbon between 1800 and 1994. "The (acidity) changes that are occurring in the oceans are truly extraordinary," said Joan Kleypas, one of the report's authors and a marine ecologist at the National Center for Atmospheric Research (NCAR).[98]

Increasing oceanic acidity has the potential to destroy two significant life forms: 1) coral and 2) tiny, shelled sea creatures. Acidity slows the growth of coral, possibly causing reefs to erode or fall apart; it also causes the carbonates used by marine life (for skeletons and shells) to dissolve. The destruction of these two living forms can significantly undermine the marine food chain that supports more complex life. Once the coral disappears, there will be no place for the tens of thousands of marine animals, plants, algae and bacteria that make a home in them; there also will be no more functioning underwater body that is capable of absorbing the wave energy generated by hurricanes and tropical storms. In turn, once the small, shelled marine creatures die, we will see effects on the more visible populations of shrimp, fish, and whales.

Skeptics of global climate change have questioned the existence of climate change and have resisted any talk of mandatory reduction for carbon dioxide emissions. Nevertheless, the relationship between rising carbon dioxide levels and the acidity of oceans is patently clear. As a recent editorial in the *Houston Chronicle* asserts, "This latest documented threat reinforces the urgent need for American industry and government to work together to curb carbon emissions created by the burning of fossil fuels. One can only hope this report will be a wake-up call that rouses our leaders to action and prevents an unthinkable future of acid oceans."[99]

After learning these facts, our team was left only with more questions. What can we do as a global community to bring about a broad-based, real resolution to this crisis? What we can we do as a global population to affect global climate change in a positive way? What can industry do to motivate the public to call for legislation that deals with climate change? I address these issues in the next chapter.

Climate change is a truly significant issue for all businesses. If a company is forward looking, if a company is focused on long-term survivability, long-term sustainability, and long-term growth, then they must factor climate change into their long-range plans.

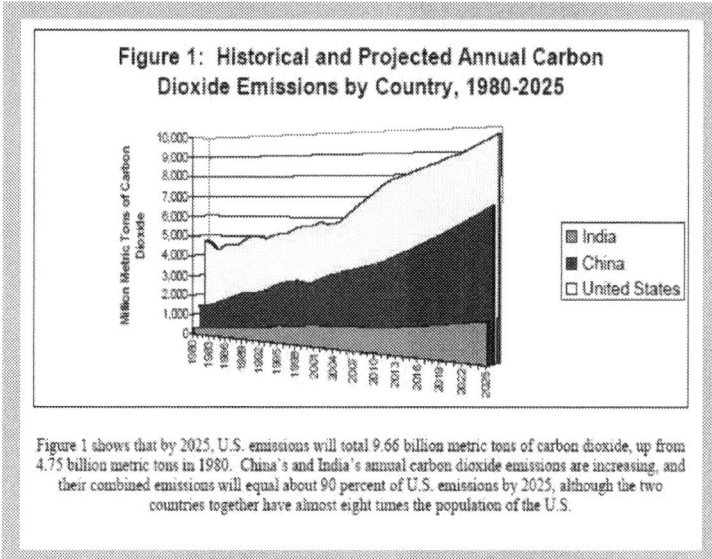

Figure 1: Historical and Projected Annual Carbon Dioxide Emissions by Country, 1980-2025

Figure 1 shows that by 2025, U.S. emissions will total 9.66 billion metric tons of carbon dioxide, up from 4.75 billion metric tons in 1980. China's and India's annual carbon dioxide emissions are increasing, and their combined emissions will equal about 90 percent of U.S. emissions by 2025, although the two countries together have almost eight times the population of the U.S.

Fig. 4-4: Annual CO2 Emissions by Country

Is it possible to remove USA from this graph? We think it is. It is crucial that the United States, home to less than 5 percent of Earth's population but a producer of 25 percent of the world's carbon dioxide emissions, take on the challenge of going green.

But American businesses need support. We need the public to support real global solutions. And when the public and the private sector work together, legislation follows.

I'm aware that major change is something most industries are reluctant to take on. But the reality is that this train is hurtling down the tracks . . . and gathering more speed. We have to prepare, and soon, and demand actions that require total accountability.

Sun Tzu said "The difficulty lies in turning the devious into the direct, and misfortune into gain."[100] What we have here is a seemingly disastrous, complicated, multi-stakeholder issue. You've got industry, you've got the public, you've got government, you've got non-governmental organizations, and there needs to be strong leadership to bring about real solutions to this problem.

Let's all become leaders. This issue—and how we confront it—will no doubt be our legacy for generations to come.

Chapter 5

The United States needs to keep an eye on emission levels, needs to promote "green" technology, needs to prepare for the changes that are going to occur as a result of climate change.

But the United States isn't the whole of the problem.

This is a *global* issue, not just a *national* issue. And if American businesses want to do more than merely survive these changes—if they want to deal with them effectively, if they want to continue to thrive and profit in the changing world that lies before us, they need to take charge of the situation. Now. There is no economic issue worth spoiling our environment; but this is an environmental issue that must be considered carefully, or we risk spoiling our environment and our economy.

My friend Dr. Thomas once said, "I can't believe that there is a smart CEO in this country who doesn't see this coming."[101] It's true. Consequently, as I outlined in Chapter 2, a lot of American companies (and a lot of American people) are beginning to go green. As *Newsweek* reported in its July 2006 issue, "with windmills, low-energy homes, new forms of recycling and fuel-efficient cars, Americans are taking conservation into their own hands."[102] Americans aren't the only ones on the move to bring about change; as James Gustave Speth notes, "The British and other Europeans are not foolish. They see the risks of climate disruption and the commercial opportunities in developing viable responses. And they know that the costs involved—perhaps no more than the difference between a 2 percent increase in GDP in 2050 and a 3 percent increase—are a small price to pay to save the planet."[103]

Who isn't going green?

Second and third world countries, that's who—those that have been exempted from responsibility by the Kyoto Treaty.

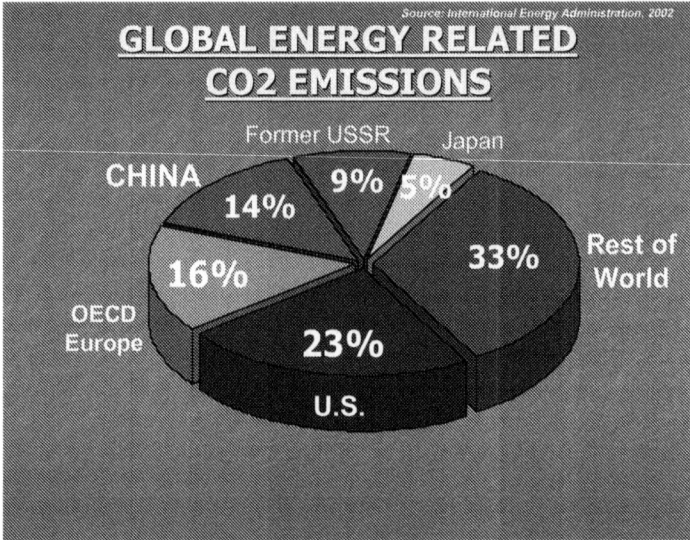

Fig. 5-1: Global Energy-Related CO2 Emissions (2002)

If you examine the pie chart above, you can see that the USA and OECD (Organization for Economic Co-operation and Development) Europe[104] are causing less than half of the emissions of carbon dioxide from energy use. Compound this with the fact that the United States and many parts of Europe are already on the move to address the problem, and one can see where focus also needs to be driven.

When we look at total greenhouse gas emissions (including all gases and all sectors), we see this pattern continued.

Developed and developing nations are almost equal in emissions based on the available data. (Annex I nations under Kyoto are generally developed nations, and required to report annually on their emissions. Although the U.S. is not a party to Kyoto, it files annual reports on its greenhouse gas emissions, as well. Non-annex I nations under Kyoto are typically "developing" nations, such as China and India. These nations are not required to report annually—in fact, China only got around to reporting its 1994 emissions levels in October 2004. Only three non-annex I nations—Mexico, Korea, and Uruguay—have bothered to report their 2000 emissions.)

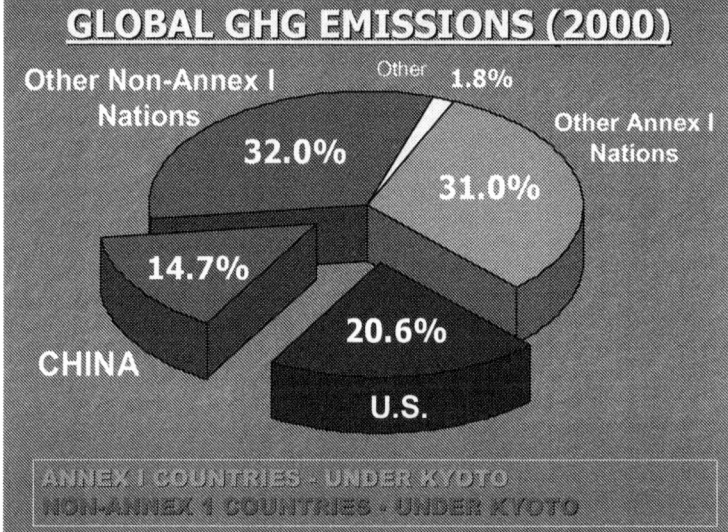

Fig. 5-2: Global Greenhouse Gas Emissions (2000)

When we include the effects of land management and carbon sinks, we see the developing world has an even greater impact on global greenhouse gas emissions—accounting for more than 57 percent of the world's greenhouse gases—and this was six years ago. Lack of timely reporting by these developing nations of their emissions makes it difficult to get an accurate picture of their true impact—but these economies have surged ahead in the past seven years, and there has been little if any concern for the environment.

Carbon sinks are carbon reservoirs that take in and store more carbon (i.e. carbon sequestration) than they release. Carbon sinks can serve to partially offset greenhouse gas emissions. Forests and oceans are large carbon sinks.

Irresponsible land use, such as clear-cutting and burning of forest, not only depletes carbon sinks, it releases decades—if not centuries—of stored carbon dioxide into the atmosphere. For example, in 2000, Indonesia and Brazil accounted for almost 50 percent of global carbon emissions from land use, such as tropical deforestation for croplands, pastures, and wood harvest. This is almost four billion metric tons of CO_2, more than the U.S. industrial and electricity generation sectors, combined.

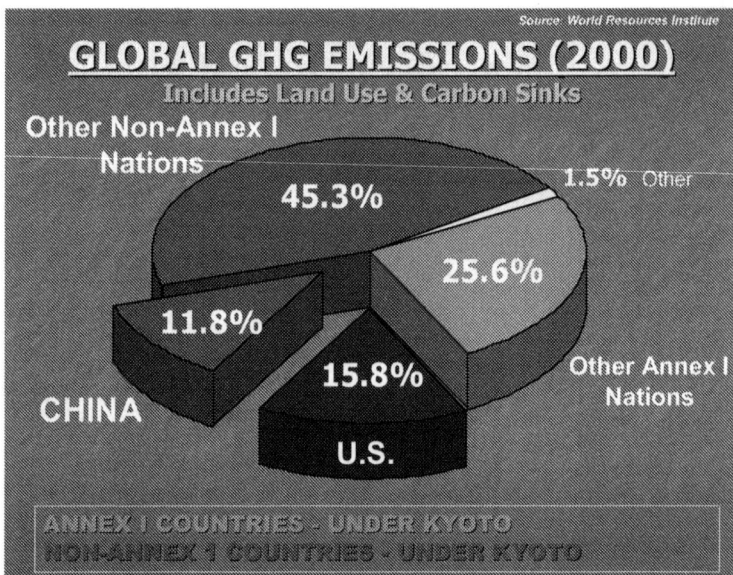

GLOBAL GHG EMISSIONS (2000)
Includes Land Use & Carbon Sinks

Other Non-Annex I Nations — 45.3%

1.5% Other

25.6%

11.8% CHINA

15.8% U.S.

Other Annex I Nations

Source: World Resources Institute

ANNEX I COUNTRIES - UNDER KYOTO
NON-ANNEX I COUNTRIES - UNDER KYOTO

Fig. 5-3: Global Greenhouse Gas Emissions (2000), Including Land Use and Carbon Sinks

Notice also that this graph does not account for where the clear cut wood is being shipped. These "developing" nations are consuming massive amounts of raw materials from around the world. So not only are they emitting greenhouse gases from their own poorly regulated industries, but they are contributing to the elimination of carbon sinks around the world.

American companies that are moving forward with renewable energy efforts, such as the invested energy from recycling, conservation, and new technologies, need support. They need to feel confident in their potential for economic growth as they make changes in the way they run, in the way they do business. Currently however, many of these companies are pinched by their need to compete with phenomenally inexpensive overseas production facilities, whose costs are kept low by largely illegal and unethical practices.

Let's talk about how this situation came about.

Kyoto started with good intentions. Leaders saw that some countries had a great deal of money and that others were mired in poverty, seemingly unable to jumpstart their economies. In order to compensate for these disparities, it was decided that the countries that weren't making any money shouldn't have to play by the same rules as the countries that were.

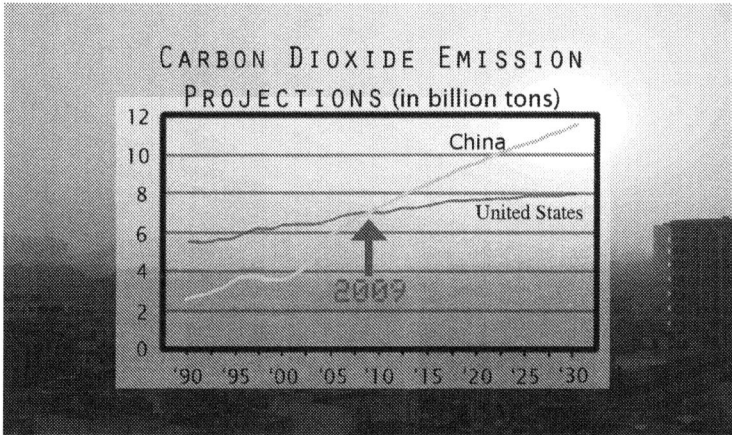

Fig. 5-4: CO2 Emissions, China and U.S.
(Source: International Energy Agency, 2006)

As the Figures 5-4 and 5-5 illustrate, Kyoto only attempts to solve a part of a very serious problem.

The consequences of Kyoto's approach to the problem?

For one, several large developing countries, including the United States, have not signed onto the Kyoto protocol.

Why?

It's because that, under the Kyoto protocol, everybody does not have to play by the same rules. As if curbing less than half of the world's greenhouse gas emissions will solve all of the world's greenhouse gas problems.

Another significant result: The "struggling" countries now pollute outrageously.

Let's take China for example.

China is considered a "developing" or "undeveloped" nation. This means it is exempt from the requirements of Kyoto. This exemption also makes China an attractive place to do business. It acts as a refuge for companies that choose to avoid production in more restrictive regulatory countries like the United States.

61

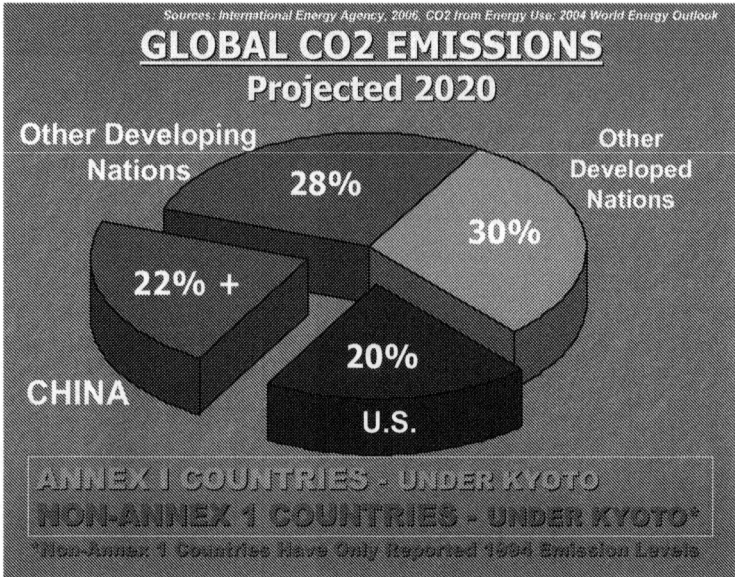

Sources: *International Energy Agency, 2006, CO2 from Energy Use; 2004 World Energy Outlook*

GLOBAL CO2 EMISSIONS
Projected 2020

Other Developing
Nations
28%

Other
Developed
Nations

30%

22% +

20%

CHINA
U.S.

ANNEX I COUNTRIES - UNDER KYOTO
NON-ANNEX 1 COUNTRIES - UNDER KYOTO*
*Non-Annex 1 Countries Have Only Reported 1994 Emission Levels

Fig. 5-5: Global CO2 Emissions (Projected 2020)

China's low environmental standards allow Chinese industries to produce an astonishing quantity of goods at low cost. But, this production—and China's surging economic development—is causing an alarming rise in its contribution to world pollution.

China is developing at a rapid pace, and so are large amounts of choking smog, poisoned rivers and toxic waste. Currently, the country is home to 20 of the world's 30 most smog-choked cities, and close to 400,000 premature deaths a year are linked to air pollution.

The Chinese government isn't doing much about all the waste. In 2005 in the Zhejiang town of Xinchang, thousands of citizens were driven to riot in order to shut down a local pharmaceutical factory that was dumping waste into the river. One news report described the situation this way:

> Villagers say they appointed a small group of rep-
> resentatives to present demands for compensation,
> including free health examinations and medical care
> for people who live near the plants, which produce a
> strain of antibiotics called quinolones. When they sent
> a group on July 4 to demand an audience with factory
> officials, they say, security guards beat the representa-

tives. The next day, the villagers returned in larger numbers and managed to grab a security officer, whom they acknowledge beating. In the meantime, as word spread of the beating of the village representatives [. . .], villagers raised the stakes, demanding the outright closing of the factory, which they had complained about for years. "Our fields won't produce grain anymore," said a 46-year-old woman who lives near the plant. "We don't dare to eat food grown from anywhere near here." Her husband, a former machine operator, said he had to quit working recently because of persistent weakness and nausea. When local officials posted a notice saying they would reopen the plant a few days after the fatal explosion there, he had been one of the first demonstrators to arrive on the scene, charging the gates and bursting into the factory with a small crowd of fellow protesters. "They are making poisonous chemicals for foreigners that the foreigners don't dare produce in their own countries," the man said. Explaining why he had been willing to rush into the plant, despite signs warning of toxic chemicals all about, he said, "It is better to die now, forcing them out, than to die of a slow suicide."[105]

In addition to the horrible civil rights violations described here, the images of the environmental pollution going on in the area are terrifying. Joshua Muldavin, a professor of human geography at Sarah Lawrence College, who has spent 20 years in rural China, sums up the development of China's current social and environmental tragedies in this way: "China's fabulous growth since the 1980s was achieved through environmental destruction and social and economic polarization which now threaten its continuation. There is an emerging pattern of rural unrest that challenges the very legitimacy of the Chinese state and the development path on which it has embarked."[106]

The Xinchang protest was one of three riots in Zhejiang over pollution in 2005. But under the country's current government, will rioting Chinese citizens successfully halt pollution?

Currently, £1.8 billion have been pledged to clean up the Songhua River on the other side of the country, poisoned in November 2005 when an explosion at a chemical plant resulted in a toxic spill that contaminated drinking water supplies for millions of people.

But will the clean-up really take place?

According to Elizabeth Economy, author of *The River Runs Black*, a book on China's battered environment, "The root cause of China's environmental problem is a fundamental unwillingness on the part of both Beijing and local governments to reform the political and economic system in a way that would make doing the right thing environmentally an attractive proposition."[107]

This is a problem.

China is playing a dangerous game. And as China's industries pollute their country, China contributes at a very alarming rate to the global climate change that threatens all of us.

Kyoto was designed to help curb the emission of greenhouse gases. However, by allowing "underdeveloped" or "developing" countries to escape environmental laws, Kyoto has brought on the development of a dangerous new world of industry, one where tons of production create tons of pollution, as Figure 5-6 illustrates, below.

Pounds of Carbon Emissions per $1,000 of GDP

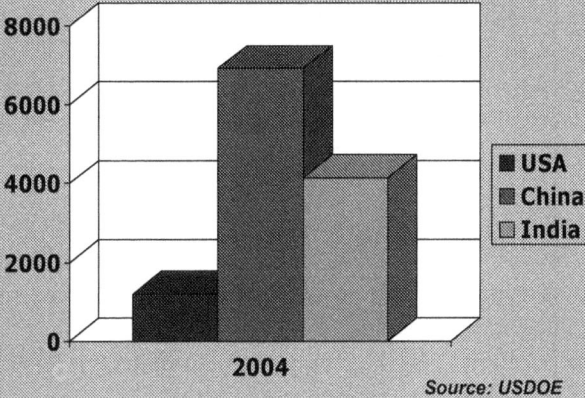

8000
6000
4000
2000
0

■ USA
■ China
□ India

2004

Source: USDOE

Fig. 5-6: Pounds of Carbon Emissions Per $1,000 of GDP

Kyoto has encouraged not the evolution of the factory system, but the movement of the factory system. Instead of revising their modes of production, companies have simply revised their location of production.

Unlike the United States, Chinese factories don't control their output of pollutants. As Steve Rowlan, General Manager of Environmental Affairs for America's Nucor Steel, notes:

> Under the Kyoto protocol, if production was transferred from the developed nations to the undeveloped nations, which would be China for example, you would have 774 percent more carbon emitted, by the tonnage of steel that was transferred to a developing nation, versus that tonnage being produced in the developed nation. We're (the United States) among the most efficient steel industries in the world. So our operations are very efficient in the amount energy we consume, in the amount of manpower we consume, the amount of emissions that are allowed to be released per ton of steel. We have strict regulations on all of that, but if you look at a developing nation, they don't.

This is a depressing truth. Are we actually encouraging carbon dioxide production—encouraging global climate change—with this treaty? This treaty that is supposed to help the world begin to address the serious problem of climate change?

That's what it looks like to me.

Ironic, isn't it? And also tragic.

Chinese cities are not the only cities affected by China's lack of environmental responsibility. Scientists have tracked chemical clouds from Southeast Asia as far as New England. China's mercury emissions are of particular concern. Mercury is a highly toxic heavy metal that damages the human nervous system. Mercury absorbed into the food we eat can cause birth defects, child development problems and cancer, and more than 53 percent of the world's mercury emissions come from Asia. The United States Environmental Protection Agency has traced high levels of mercury found in the U.S. back to Chinese facilities such as unfiltered coal-fired electric plants.

To put it mildly, more than two decades of 9.5 per cent annual growth in China's economy has come at a cost.

We're all going to feel the results of this waste, especially because, as China continues to pollute, it also continues to acquire business.

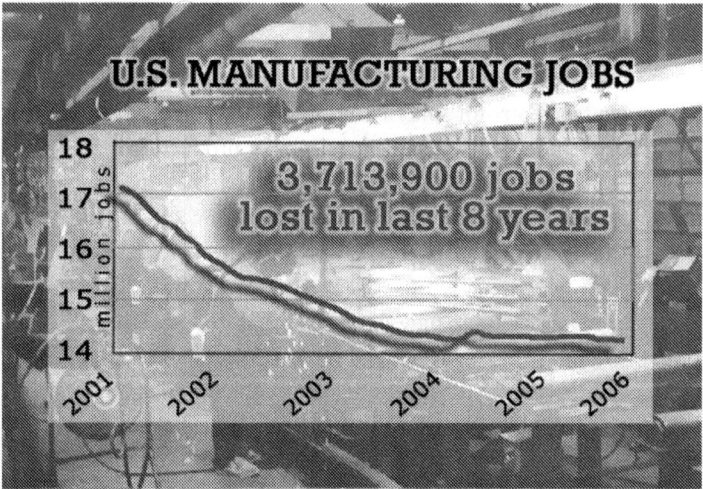

Fig. 5-7: U.S. Manufacturing Jobs (Source: U.S. BLS)

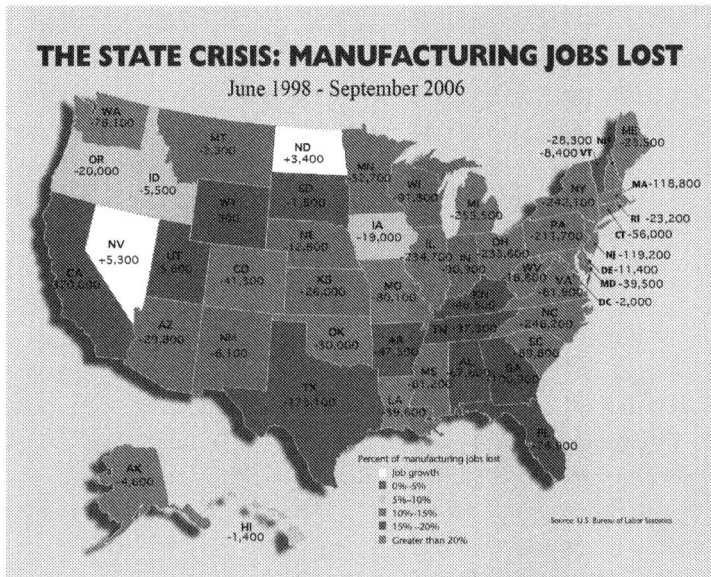

THE STATE CRISIS: MANUFACTURING JOBS LOST
June 1998 - September 2006

Fig. 5-8: U.S. Manufacturing Jobs Lost
(June 1998—Sept. 2006, U.S. BLS)

In January 2005, a study sponsored by the United States' Economic Policy Institute found that our growing trade deficit with China has had an increasingly negative impact on our economy. The study, titled "U.S.-China Trade, 1989-2003: Impact on Jobs and Industries, Nationally and State-by-State," found job losses that extend into the most technologically advanced industries in the manufacturing sector and that affect every state.[108]

Dr. Robert Scott, who helped organize the study, notes that "The assumptions we built our trade relationship with China on have proven to be a house of cards. Everyone knew we would lose jobs in labor-intensive industries like textiles and apparel, but we thought we could hold our own in the capital-intensive, high-tech arena. The numbers we're seeing now put the lie to that hope—as China expands its share even in core industries such as autos and aerospace."[109]

Specifically, the study found that:

- The rise in the United States' trade deficit with China from 1989 to 2003 caused displacement of production that supported 1.5 million U.S. jobs. The loss of jobs due to the growing trade deficit with China has more than doubled since China entered the WTO in 2001.

- China's exports to the United States of electronics, computers, and communications equipment, along with other products that use more highly skilled labor and advanced technology, are growing much faster than its exports of low-value, labor-intensive items such as apparel, shoes and plastic products.
- The U.S. trade deficit in Advanced Technology Products (ATP) with China is now $32 billion, equal to the total U.S. ATP deficit with all other countries combined.
- China is also rapidly gaining advantage in more advanced industries such as autos and aerospace products.

Regarding the study, United States-China Economic and Security Review Commission chairman C. Richard D'Amato commented, "With such data, we can begin to assess the impacts China is having on our economic health and our national security."[110]

Yes, with such data we can. And we should.

In my opinion, the way to describe it is just exactly like it is. What we're seeing around the world is just plain old-fashioned cheating.

Let me explain.

Certain foreign governments *manipulate* their currencies to ensure favorable exchange rates against the U.S. dollar and ensure that products manufactured in their nations have an illegal advantage—as much as a 40% discount.

Let's get specific. For more than a decade, China "pegged" its currency to the U.S. dollar, making its own currency artificially undervalued. Although China declared just last year that it had made plans to allow its currency to "float" against a "basket" of currencies, there has been no significant change in the way that they value their currency. China has limited the fluctuation in its currency, and China's central bank still intervenes in the currency market.

By manipulating its currency, China has placed an obstacle in the way of free and fair global markets.

By failing to revalue its currency, China continues to contribute to the long-standing currency imbalances that have harmed our economy for decades. Although China recently did claim to cut back its peg of the yuan to the dollar, the actual size of this move, as illustrated by the chart below, has been ridiculously insignificant.

PACIFIC EXCHANGE RATE SERVICE
Monthly Average Exchange Rates: U.S. Dollars per Chinese Renminbi

0.18
0.17
0.16
0.15
0.14
0.13
0.12
0.11

1995   2000   2005

Fig. 5-9: Monthly Average Exchange Rate—U.S. Dollars per Chinese
Renminbi (1995-2005)

Some U.S. policymakers have charged that China's currency is undervalued by as much as 40 percent in relation to the U.S. dollar, and academic studies support this contention. Also, China has failed to discontinue its purchase of large amounts of dollars, which further contributes to currency and trade imbalances. Joining China in this practice are the other East Asian U.S. trading "partners" Japan, Korea and Taiwan.

The consequences of all this? Chinese exports to U.S. markets are unfairly cheap while U.S. exports to China are unfairly expensive.

It is virtually impossible for U.S. goods to compete on price in global markets.

This inability to compete results in drastically reduced demand for U.S. goods.

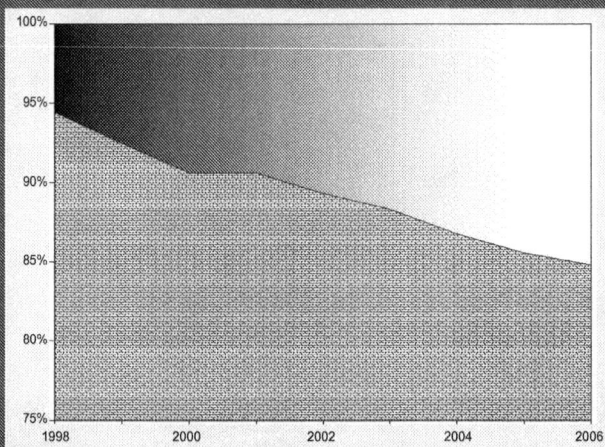

Domestic Manufacturers are Steadily Losing Market Share in the United States

Source: U.S. Census Bureau

Fig. 5-10: Domestic Manufacturers Share of U.S. Market (1998-2006)

U.S. policymakers now argue that China's practice of undervaluing its currency has directly contributed to the burgeoning U.S. trade deficit with China described above. It is estimated that as much as one-fourth of the United States' trade deficit with China is a direct result of foreign currency manipulation! While currency manipulation is certainly not the sole cause of the overvaluing of the dollar, the practice keeps the value of U.S. currency inflated while preventing foreign currency from appreciating.

The International Monetary Fund (IMF) Articles of Agreement describes the purpose of the international monetary system as, "a framework that facilitates the exchange of goods, services, and capital among countries, and that sustains sound economic growth."[111] The IMF specifically addresses the illegality of currency manipulation by stating that governments should, "avoid manipulating exchange rates or the international monetary system in order to prevent effective balance of payments adjustment or to gain an unfair competitive advantage over other members."[112]

Yet policymakers in D.C. hide behind notions of "free trade" when refusing to intervene to stop this illegal practice. This is an outrageous abuse of the principles of free trade. Free trade is not "free-for-all" trade. Free trade is supposed to be based on the rule of law and equitable principles established—and enforced—under treaties and international laws. Laws that are routinely broken by our trading "partners." Stakeholders who seek enforcement of these laws are not opponents of free trade or "protectionists"—they are seeking to uphold the equal treatment of United States and foreign businesses under the laws to which we have all agreed.

Consequently, American industry find itself unable to realize the benefits of its increased capacity and productivity.

The loss of jobs to China is certainly bad news for American workers, but because of the irresponsible way that the Chinese regulate their means of production it's worse news for our environment. The Chinese and other "developing" nations are besmirching their (and our) environment at alarming rates.

The Chinese want to develop as a nation. But they shouldn't be able develop by stealing jobs away from citizens of the United States.

As Paul Craig Roberts, Assistant Secretary of the Treasury in the Reagan administration, Associate Editor of the *Wall Street Journal* editorial page and Contributing Editor of *National Review*, asserts:

> The export of high-productivity jobs means that foreign nationals abroad earn the incomes from producing goods sold by U.S. companies in U.S. markets. Policy-makers and trade enthusiasts assume that this is a rosy development, because it means lower prices for consumers and holds down inflation. They ignore the other side of the equation: lower growth in per capita incomes from the loss of high-productivity jobs.[113]

As jobs move to China, America's unemployment rates increase, as does consumer debt.

The consumer sector accounts for two-thirds of the economy, yet it is deeply in debt. How can our economy ever recover? Disturbingly, Roberts notes that "The income growth needed to drive the U.S. economy cannot come from exporting manufacturing jobs and replacing the jobs with retail clerk jobs selling foreign-made goods."[114]

Fig. 5-10: Total Consumer Debt
as Percent of Disposable Income

In his article, "A Nation of Waitresses and Bartenders?" Roberts cites America's dismal history of lost jobs to China over the past six years:

> During 2001 and 2002 the U.S. economy lost 2,298,000 jobs. These lost jobs were not regained until early in February 2005. From February 2005 through April 2006, the economy has gained 2,584 jobs (mainly in domestic services).
>
> The total job gain for the 64 month period from January 2001 through April 2006 is 7,000,000 jobs less than the 9,600,000 jobs necessary to stay even with population growth during that period. The unemployment rate is low because millions of discouraged workers have dropped out of the work force and are not counted as unemployed.

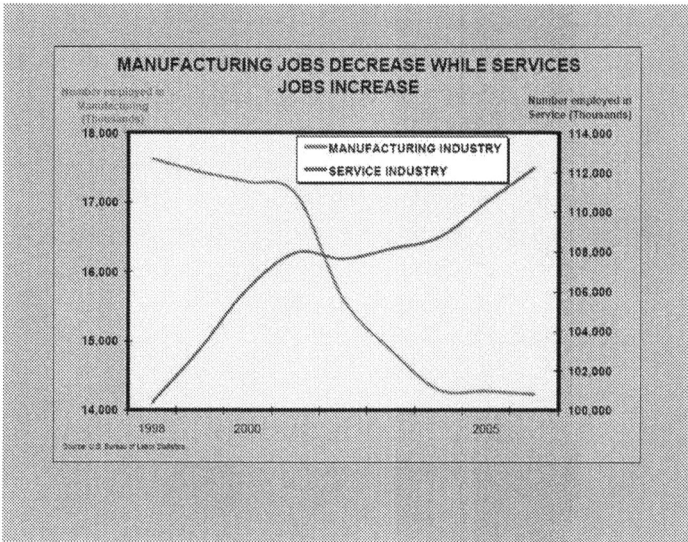

Fig. 5-11: Manufacturing and Service Jobs (1998-2006)

In 2005 the U.S. had a current account deficit in excess of $800 billion. That means Americans consumed $800 billion more goods and services than they produced. A significant percentage of this figure is offshore production by U.S. companies for American markets.

The U.S. current account deficit as a percent of Gross Domestic Product is unprecedented. As more jobs and manufacturing are moved offshore, Americans become more dependent on foreign made goods. This year the deficit could reach $1 trillion.

The U.S. pays its current account deficit by giving up ownership of its existing assets or wealth. Foreigners don't simply hold the $800 billion in cash. They use it to acquire U.S. equities, real estate, bonds, and entire companies.

The federal budget is also in the red to the tune of about $400 billion. As Americans have ceased to save, the federal government is dependent on foreigners to lend it the money to operate and to wage war in the Middle East.

73

American consumers are heavily indebted. The growth of consumer debt is what has been fueling the economy. Social Security and Medicare are in financial trouble, as are many company pension plans. Decide for yourself -- is this the economic picture of a superpower that can dictate to the world, or is it the picture of a second-rate country dependent on foreigners to finance its consumption and the operation of its government?[115]

A frightening question, isn't it?

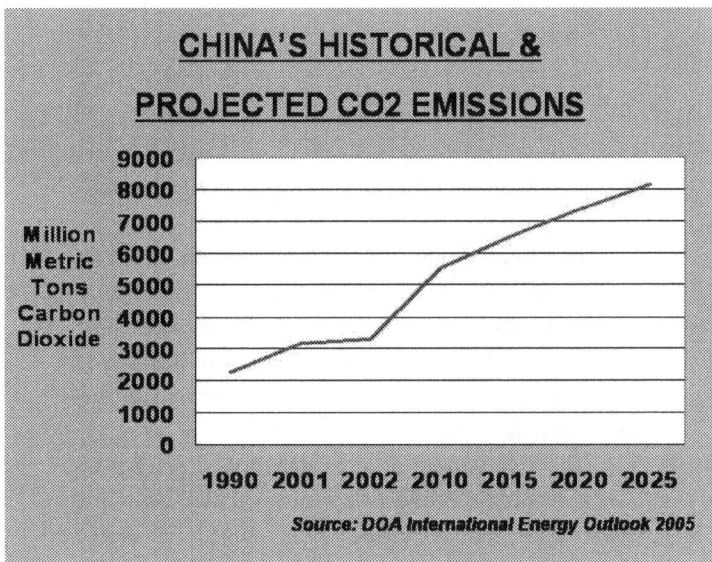

Fig. 5-12: China's Historical and Projected CO2 Emissions

The road to Chinese prosperity is not going to be built on the back of the American worker. And I feel the same way about our environment. Our planet is now being despoiled at a much higher rate because we are failing to enforce any sort of global production standards.

As China acquires more business, their carbon dioxide emissions will continue to rise . . . and rise . . . and rise.

Nucor Steel's General Manager for Environmental Compliance, Steve Rowlan, notes:

> In my opinion, they've really got Kyoto backwards. If you were to take a reduction in production in the developed nations and shift it to a non-developed nation, the amount of carbon emissions actually goes up. The net is an actual loss to the environment, while the developed nation has a loss of jobs associated with its contribution to Kyoto. Now we start to see because of unfair currency manipulation, trade policy, through just the cost of production, we see industry leave the United States, and go over seas, and as I've said before, $1000 of GDP shift to China is 774% more pollutants being discharged than it would be in the U.S.[116]

Nucor Steel, while only one of the companies in the United States that is thinking about how it can provide a quality product, made by Americans and through a process that does not contribute to the destruction of our planet, really illustrates the productive progress a concerned corporation can make when faced with new environmental standards. Rowlan explains further:

> [W]hen you look at the U.S. Iron and Steel sector, as of 2003, using U.S. EPA numbers, our emissions—our carbon related emissions—are 33% below 1990 levels. That's roughly 6-7 times what Kyoto could have expected of us, and that's been achieved through some realignment in the industry, by making the industry much more efficient, and removing the inefficient players from the industry. Now many would say that that's business as usual. If that's business as usual, and that's the kind of results that they get, then I think that's what Kyoto ought to be asking for.
>
> The prudent company, and what we at Nucor believe we are obligated to do, is to look at that and ask how do I stay in front of these regulations, how do I strategically position myself to take advantage of opportunities that may come out of this. The company that isn't strategically thinking ahead [. . .] is going to

worry about them when the rules come down. At Nucor, we've tried to not allow ourselves to get caught in the reactive mode.[117]

The US Iron & Steel Industry has already bettered Kyoto by 700%

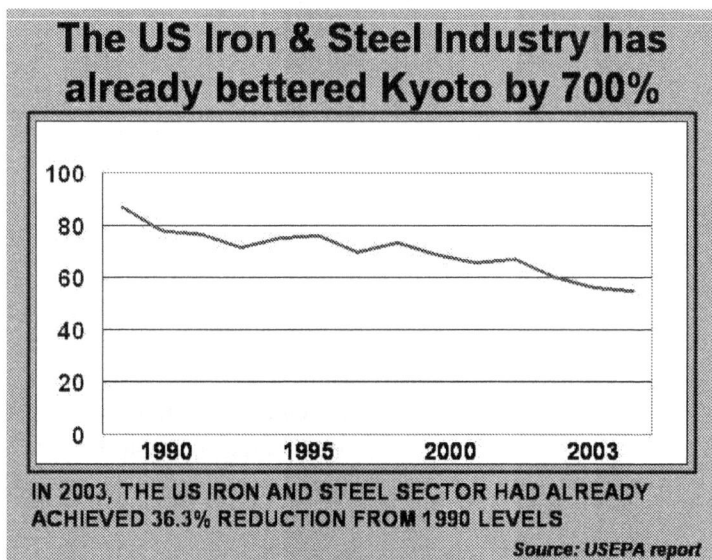

IN 2003, THE US IRON AND STEEL SECTOR HAD ALREADY ACHIEVED 36.3% REDUCTION FROM 1990 LEVELS

Source: USEPA report

Fig. 5-13: CO_2 Emissions from U.S. Iron and Steel Sector

Not surprisingly, Nucor, the largest steel recycler in the United States, has been recognized as a leader in environmental stewardship. Despite this amazing progress, Nucor still has to fight for a level playing field.

And it's not just China they're competing against. Other polluting countries are manufacturing more every day, often to the harm of our planet.

India's rapidly growing population has placed a significant stress on the country's natural resources. Deforestation, soil erosion, water pollution and land degradation continue to worsen in rural India, while urbanization in India has resulted in serious air pollution problems. Indeed, India has more than 20 cities with populations of at least 1 million, and New Delhi, Mumbai, Chennai, and Kolkata are among the most polluted cities in the world. Urban air quality in these cities is especially bad. Airborne particulate matter (PM) has been registered at levels more than 10 times India's legal limit.

According to America's Energy Information Administration:

Of the 3 million premature deaths in the world that occur each year due to outdoor and indoor air pollution, the highest number are assessed to occur in India. Sources of air pollution, India's most severe environmental problem, come in several forms, including vehicular emissions and untreated industrial smoke. Continued urbanization has exacerbated the problem of rapid industrialization, as more and more people are adversely affected and cities are unable to implement adequate pollution control mechanisms.[118]

And this is the environmental state of the country that is taking jobs away from American corporations?

India signed the United Nations Framework Convention on Climate Change as a non-Annex I country in 1992. By doing this, the country was designated free from any responsibility to reduce its emissions of carbon and greenhouse gases. Because the Indian government places a high priority on economic development, the country chose not to participate in the Kyoto Protocol.

According to America's Energy Information Administration,

In 2001, India, with 251 million metric tons of carbon equivalent emitted, ranked fifth in the world in carbon emissions, behind the United States, China, Russia and Japan. Although India's carbon emissions stood at only 80% of Japan's (316 million metric tons of carbon equivalent) total and less than one-sixth of the United States' (1,565 million metric tons) carbon emissions that same year, the rapid growth of India's carbon emissions—in combination with its exclusion from the Protocol—is [a] main point of controversy. Between 1990 and 2001, India's carbon emissions increased by an astonishing 61%, a rate surpassed only by China's 111% increase during the same time period. India's carbon emissions are expected to continue to increase throughout the decade, offsetting the planned reduction in GHGs from the European Union and other countries that plan to implement the Protocol's emissions cut requirements. The rise in India's carbon emissions has been exacerbated by the low energy efficiency of coal-fired power plants in the country. With the high capital costs associated with replacing existing coal-fired plants, a scarcity of capital, and the long lead time required to introduce advanced coal technologies, it stands to reason that many of India's highly-polluting coal-fired power plants will have to remain in operation for the next couple of decades. As such, India's contribution to world carbon emissions is expected to increase in coming years, with an estimated average annual growth rate between 2001 and 2025 of 3.0% in the EIA International Energy Outlook 2003 reference case (compared to 3.4% in China and 1.5% in the United States). [...] [T]he country's per capita carbon emissions are expected to increase in the coming years due to the rapid pace of urbanization, a conversion away from noncommercial towards commercial fuels, increased vehicular usage and the continued use of older and more inefficient coal-fired plants. In fact, due to fast-paced industrialization, per capita emissions are expected to triple by 2020.[119]

Such emissions are destined to cancel out the environmental efforts to curb carbon dioxide production that are being made in the developed countries, such as the United States.

China, India, Mexico, Brazil and Argentina ... American jobs are moving to countries where labor is underpaid, where energy consumption and consequent carbon dioxide emissions are skyrocketing, and where rapid urbanization has intensified the problem of rapid industrialization, to the point where cities do not employ adequate pollution control mechanisms. These are the reasons why it is "cheap" to export manufacturing jobs overseas.

The real issue that drives America's economic woes is the fact that we are competing against countries that are cheating. And the real basic issue that drives America's environmental woes is the fact that we can't clean up the environment when half of the world's polluters are not going to even try to help.

Not only will they not try to help, they'll do the very things we're trying to prevent.

And that's where we stand today.

Pollution is a worldwide problem. All industries, in all countries, need to take responsibility for our planet, and to be held accountable for their treatment of the environment. It's time that everyone begins to understand just how dangerous our climate situation is. It's also time that we begin to understand how important it is that societies and businesses and governments and scientists all come together to try to take what preventive measures we can.

I know my ideas may seem difficult to put into operation, but global climate change is a problem for the whole world. We need to understand that it will take all of us working together to begin to do something about it.

One current impediment to change is our government. It is imperative that we come up with a plan of action that does what we hoped Kyoto would do. Obviously, we need to be aware of the ways in which our own country might be contributing to the problem; however, we also need to use our leverage as the world's most powerful economy. We need to make sure that other countries are living up to the agreements they've signed about trade and about the environment.

What else can be done?

Assessment. Alignment. Concurrence.

Who are the stakeholders? What is at stake? How can different stakeholders work together to bring about mutually beneficial solutions?

It's time to raise awareness. Once that is done, the American people can voice their needs—for jobs and for safe places to raise healthy families; businesses can voice their needs—for a stable consumer base, for the money to work towards more efficient, less ecologically unfriendly production methods; and governments can begin to put their practices in line with those needs.

It's time to face the music. We should invest in the future health of our planet. We should put tax credits, grants, and incentives towards companies who make green choices. We should make the public aware of where and how the everyday products that they use are produced.

It's easy to draw a line and take a dogmatic, philosophical position that either industry should stop or that change is too costly on industry. Although both of those positions are currently being voiced, although both of these positions are important in driving the debate—at the end of the day they are both losing positions.

There is no amount of money that's worth spoiling our environment over. At the same time, there's no one single issue that, absent legitimate consideration, is worth spoiling our economy and our workplace for our working families. So in many ways, what we hope to come out of this is a very balanced, yet real, proposed solution.

We need a solution that goes beyond either "You need to sign Kyoto" or "You don't," or "It costs too much" or "Costs shouldn't matter"—to really dealing with this with a non-zero sum approach. I learned this long ago on the streets of South Central Los Angeles. Solutions are complicated. They involve all sorts of people working together. They involve sacrifices. They involve the challenges of rethinking our settled beliefs and our habitual ways of thinking. They also involve rewards. Through innovation and discipline, we can work to create a healthier planet. At the same time, we can also work towards a healthier American economy.

Talking about a proposed solution envelops all stakeholders, and this process is one that I'm eager to begin. My team and I, we're really all about kicking the tires. I say you don't do politics over the telephone. You don't deal with environmental remediation over the telephone or read about it in the books. You get down, roll your sleeves up first hand, and take a look—that's what my process is all about.

Theodore Roosevelt once said: "Far and away the best prize that life offers is the chance to work hard at work worth doing." That's why we went to Antarctica.

80

Afterword

Since my journey to Antarctica, it has become evident that federal, state, and local governments are preparing to take significant action on global climate change in the United States. That is why we will continue to engage this issue by venturing to areas where the impact of global climate change can be studied, first hand. Those who are trying to manage this issue from climate controlled offices in Washington, D.C. simply cannot comprehend the importance of taking meaningful action in order to effect real change. If those in power are permitted to treat climate change as a political problem, it will be handed down to our children as a real problem that they will be forced to find real solutions to, and our generation's legacy will be one of failure to take responsibility.

In Antarctica, any effects of climate change stand out against the stark backdrop of an uninhabited land. But in other places around the world, climate change has already become much more than a matter of science. In places like Sub-Saharan Africa, the changes in the environment aren't just being witnessed by researchers—they are being felt by the people who inhabit and depend upon the land.

In October 2006, we organized a second expedition to Namibia, Africa, in order to witness and study the threat that climate change poses to the global community. The findings and success of our Antarctica trip allowed for this second portion of the symposium to take place. This time we set out to examine the challenges posed by climate change to human existence and to further explore meaningful and responsible solutions. Again, we brought together a multi-disciplinary team of professionals to explore these important issues.

Sub-Saharan Africa has some of the most fragile ecosystems in the world. Namibia is expected to face a water scarcity by 2020. Ranked among countries that are likely to be the most vulnerable to climate change in the coming years, Namibia's water sector will be the hardest hit. Much like Antarctica, Namibia faces potentially catastrophic effects from climate change, most notable along the coastline and in the arid Namib Desert. Beyond the more obvious ramifications of lower annual rainfall, however, equally ominous effects await the citizens of this country.

As a result of global climate change, weather events will likely become less predictable and more intense, while heavier rainfall and longer and more frequent drought cycles will persist. Human health will probably be similarly affected, as otherwise treatable diseases and conditions become exacerbated by the lack of water. Insects and vermin would thrive, leading to widespread disease. For example, small changes in temperature and precipitation will boost population of disease-carrying mosquitoes and result in increased malaria epidemics. Increased flooding can facilitate the breeding of these malaria carriers in formerly arid areas.

The main causes of death in Namibian children under the age of 5 all have environmental components. Drought decreases nutritional status, particularly in rural areas, and reduces the availability of water. As one of the driest countries in sub-Saharan Africa, Namibia will be hit hard by the effects of climate change, says Environment and Tourism Minister Willem Konjore. "The effect of climate change will drastically increase the vulnerability of people within southern Africa and Namibia, which is already compounded by the threats of desertification," Konjore said.

Namibia is vulnerable because of its fragile ecosystems and a subsistence economy that relies heavily on subsistence farming. Sporadic droughts have caused considerable stock losses and reduced grain production, severely affecting the poorest farmers and rural communities. As a result of global climate change, sea levels along the Namibian coast are expected to rise between 30 and 100 centimeters by the year 2100, resulting in the warming of the cold Benguela Current.

As annual rainfall decreases in this southern African nation, the world's most precious resource could very well become the subject of regional conflicts. An increase in violent conflict and a refugee crisis will likely result from the impact of global climate change on water supplies in poor countries. "Even in the case of a moderate increase in evaporation of 15 per cent and no change in rainfall, the additional stress on the water sector due to climate change would be severe," says Namibia's report to the United Nations Framework Convention on Climate Change.

Climate change may sow the seeds of violent conflict, as people battle for control of increasingly scarce resources. An example of where this catalyst for conflict has already begun on the African continent can be seen in the political wrangling among Namibia and neighboring Botswana and Angola. The three countries have been at loggerheads over the Okavango River Basin. Tensions arose when Namibia proposed to pipe water from the river to its capital amid worsening droughts.

Even more disturbing than the impact and potential impact of climate change on the Namibian people is the growing influence of China on the African region. Sub-Saharan Africa has abundant natural resources, and China has been aggressively establishing itself as a major consumer of Africa's timber, oil, and minerals. While China's involvement in Africa may have some short-term economic benefits for an elite group of Africans, their activities threaten to have devastating long-term environmental consequences for Africa and the world. China is acting with even less environmental responsibility in Africa as it has within its own borders—a frightening thought.

China's interest in Africa is being fueled by its rapid economic expansion. China is quickly becoming the world's largest consumer of oil. By 2020, China will surpass the United States in terms of the number of cars on the roads and daily oil consumption. To meet its growing demand, China has started investing heavily on African oil rights. A Chinese company recently paid $1 billion for exploration rights off the coast of Angola. Similar oil prospects off of Angola were selling for $35 million less than a decade ago.

The Chinese government has been involved in a bartering system with African countries to acquire exploration rights for less than market value. The Chinese government recently agreed to build a $500 million dam in Ghana, a country that is receiving $8 billion in debt relief. The write-off would leave the country with a debt-to-export ratio of just 30 percent. But new borrowing from China and elsewhere is likely to drive that back up to 150 percent before the first decade of the 21st century is over.

China is currently in the business of locking up energy supplies and is doing so by exploiting African geopolitical instability. It has supported governments notorious for dictatorial and harsh regimes. It has overpaid for energy exploration rights to ensure exclusivity, indicating a massive buildup of industrialization, further pouring toxic pollutants into the environment.

China's environmental record is nothing short of abysmal. Pollution from burning the country's highly sulfurous coal, China's leading source of energy, is most damaging to the health of Chinese, the quality of the air they breathe and the water they drink. Half of the 2 billion tons of coal China burns each year, accounting for two-thirds of the country's energy consumption, feeds industry. The sulfur dioxide and soot thrown off results not only in a polluted atmosphere but also acid rain that falls on a third of China's cities.

By 2020, China is projected to account for 22 percent of the world's greenhouse gas emissions. According to the Asia Pacific Partnership on Clean Development and Climate, China and India are projected to account for more than 37 percent of the world's greenhouse gas emissions by 2050, while the U.S. will have reduced its share to 15 percent. If the United States acts only to further reduce its own emissions, it will simply exacerbate the pollution problems in China by fueling further economic expansion—and environmental abuse—in nations such as China.

As we convened in the Namib Desert for our symposium, we considered as a group the validity and implications of global climate change. As with our previous expedition to Antarctica, we were struck by a conflicting set of seemingly polarized conclusions: the vastness of our global environment, and simultaneously the inter-related system of our global community. Peeling back the various layers of policy debates surrounding global climate change, we found the issues and concerns of the individuals and communities amazingly similar.

It is almost inconceivable to me that the best thoughts that we can produce in our global community to address the real and immediate challenges posed by our energy and environmental concerns, centers on a document (the Kyoto Treaty) so flawed and counterintuitive as almost to defy logic. And yet, many of us are moved to accept proposed solutions that are equally meaningless.

Like a star soccer or basketball player who, sensing his team is perilously behind, is going to pull the team forward by himself, we see individual states in the United States attempting to go it alone. But just as in any team sport, the results will no doubt be equally fruitless. This is not leadership, but rather leading with your chin.

I return again to Dean Gus Speth's statement from *Red Sky at Morning* that we must look beyond our own borders for solutions:

> [O]ur national legislation was successful in curbing many environmental abuses domestically... [but]

more of the same will not get us where we want to be in time to head off an era of unprecedented environmental decline. . . . The current system of international efforts to help the environment simply isn't working. The design makes sure it won't work, and the statistics keep getting worse. We need a new design, and to make that happen, civil society must take the helm.[120]

Failing to heed this insightful advice, Congress and some states are considering mandatory twenty-five percent reductions in emissions from utilities and/or industry. These reduction programs—if successful—will have a marginal to *de minimis* effect on global emission reductions, and will no doubt hasten further departure of industry from the United States to nations where the emissions levels will be many times greater. Take for example, the recent legislation passed in California. If the state manages to meet it's goal of an approximately 25% reduction in greenhouse gas emissions, it will only result in a 0.37% reduction in global emissions—assuming all else remains equal. However, the actual effect on global emissions may be worse. The increased regulatory costs will lead to further erosion of the domestic manufacturing base. As the U.S. Bureau of Labor Statistics has reported, California has lost almost 370,000 manufacturing jobs in the last decade. California also has some of the most progressive and aggressive air quality standards in the world. There is no reason to think that further, unilateral regulations will not expedite additional movement of industry to developing nations.

What we need is a global solution for a global issue. We simply cannot afford to squander the energy and attention given to this issue by failed local solutions. Failure to seek a meaningful global solution could have disastrous environmental and economic consequences on our great nation, and the world—the perfect storm.

Global Solutions

Demanding Total Accountability for Climate Change

Antarctica has experienced significant retreat and collapse of ice shelves, which has been the result of regional warming.

The Larsen B ice shelf, the size of Rhode Island, had been stable for 10,000 years. In 2002, it broke apart in a matter of months.

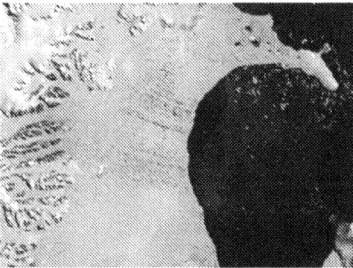
January 31, 2002

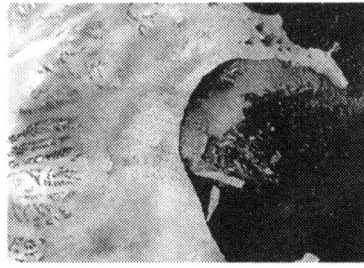
February 17, 2002

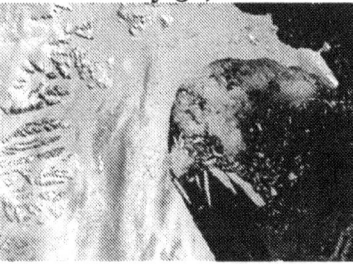
February 23, 2002

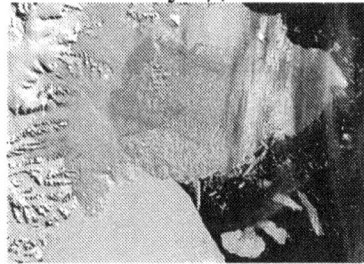
March 5, 2002

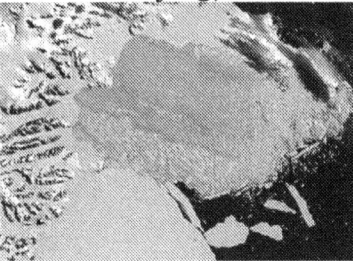
March 7, 2002

Courtesy of NASA and Visible Earth
(http://visibleearth.nasa.gov/)

The Kyoto Treaty was the first attempt at an international solution to climate change.

Unfortunately, Kyoto exempts "developing countries" from emission reduction requirements.

"Developing countries," like China, represent an ever-increasing percentage of greenhouse gas emissions.

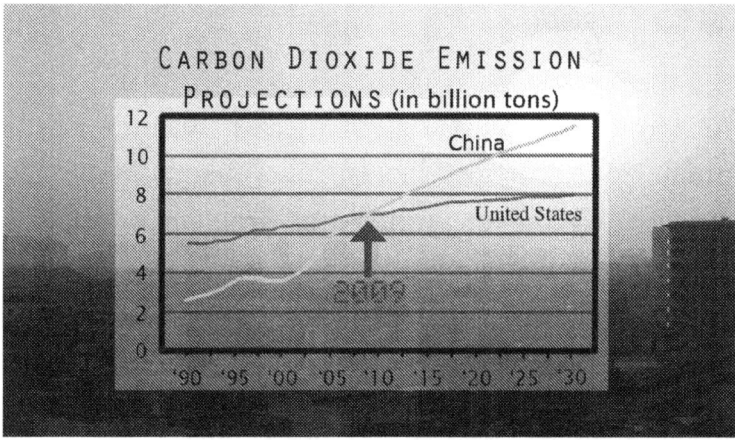

CARBON DIOXIDE EMISSION PROJECTIONS (in billion tons)

China's emissions have grown more in a year than U.S. emissions have in a decade.

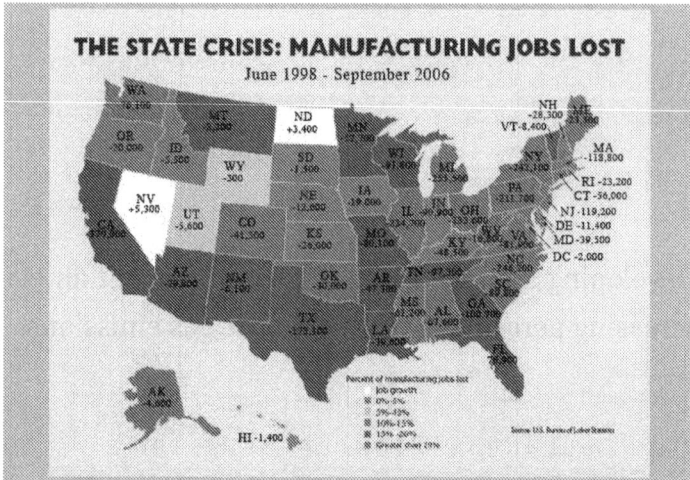

THE STATE CRISIS: MANUFACTURING JOBS LOST
June 1998 - September 2006

American manufacturing job loss resulting from unfair trade practices, like currency manipulation, has exacerbated the environmental problem by driving manufacturing to nations without any discernible environmental regulations.

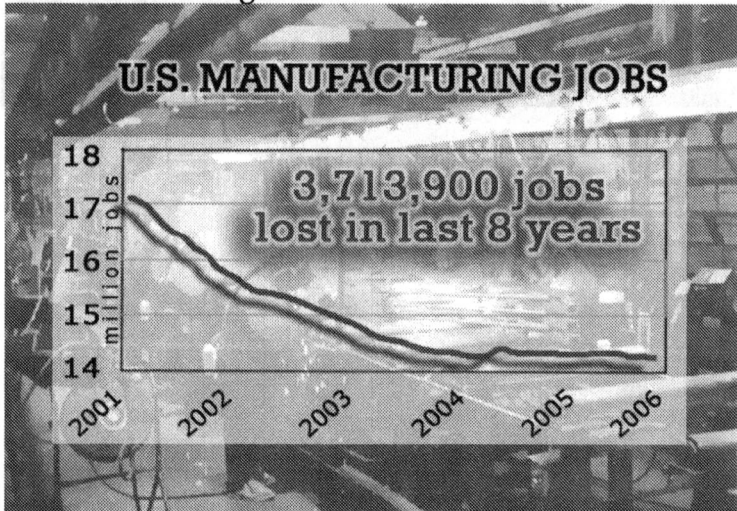

U.S. MANUFACTURING JOBS

3,713,900 jobs lost in last 8 years

China has 7 of the 10 world's most polluted cities and is building the equivalent of two 600 megawatt coal fired power plants a week to support their burgeoning economy.

Increasing reports of global climate change are spurring action. Some states, such as California, have decided to take unilateral action. Unfortunately, such action promises little reduction in global greenhouse gas emissions.

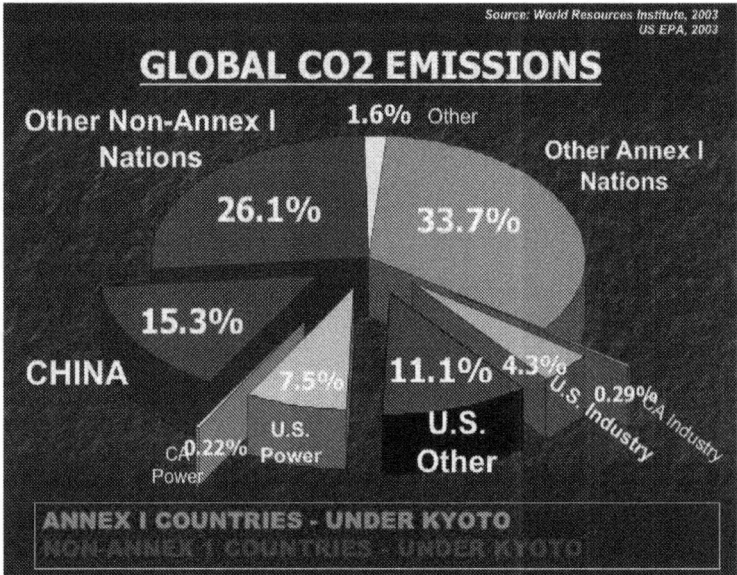

Even if California managed to reach its stated goal of reducing its industrial and power utility emissions by 25% by 2020, it would account for only a 0.13% reduction in global emissions; and that's assuming that these reductions did not lead to an increase in emissions as industry and utilities moved to other states or nations with less stringent emissions standards.

Environmental regulations, such as cap-and–trade, that further drive manufacturing to counties without equivalent regulations threaten an environmental "backfire" by increasing overall greenhouse gas emissions.

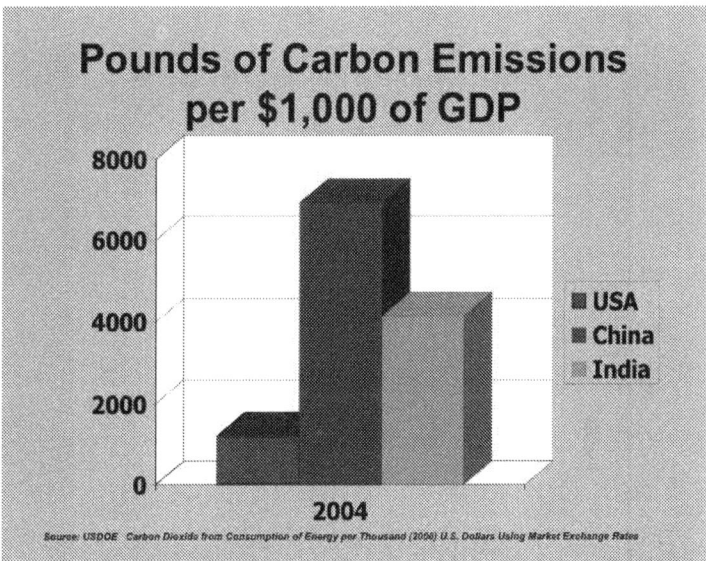

Pounds of Carbon Emissions per $1,000 of GDP

Source: USDOE Carbon Dioxide from Consumption of Energy per Thousand (2004) U.S. Dollars Using Market Exchange Rates

Every $1,000 of production shifted overseas can lead to up to five times the amount of carbon dioxide being emitted into the atmosphere.

Cap-and-trade regulations also threaten to punish U.S. industry that have been environmentally proactive. For instance, the U.S. steel industry has already reduced its emissions 37% below 1990 levels.

The US Iron & Steel Industry has already bettered Kyoto by 700%

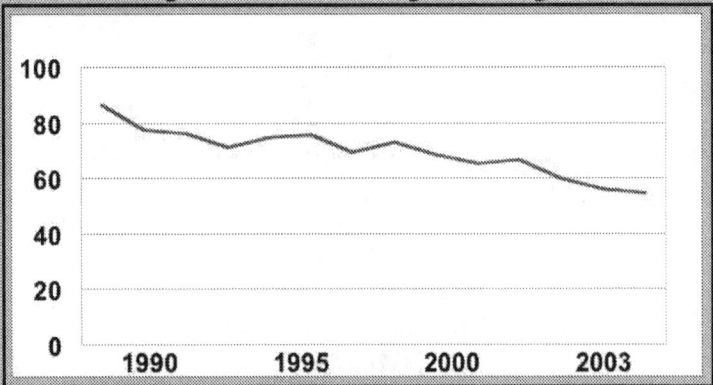

IN 2003, THE US IRON AND STEEL SECTOR HAD ALREADY ACHIEVED 36.3% REDUCTION FROM 1990 LEVELS

Source: USEPA report

Responsible solutions to climate change must provide total accountability.

A "carbon tax" would require comparable reductions in GHGs by other nations and would limit the "migration" of emissions.

Greenhouse gas emissions don't just come from industrial sources. We must also encourage improvements in the energy efficiency of the American economy across all sectors.

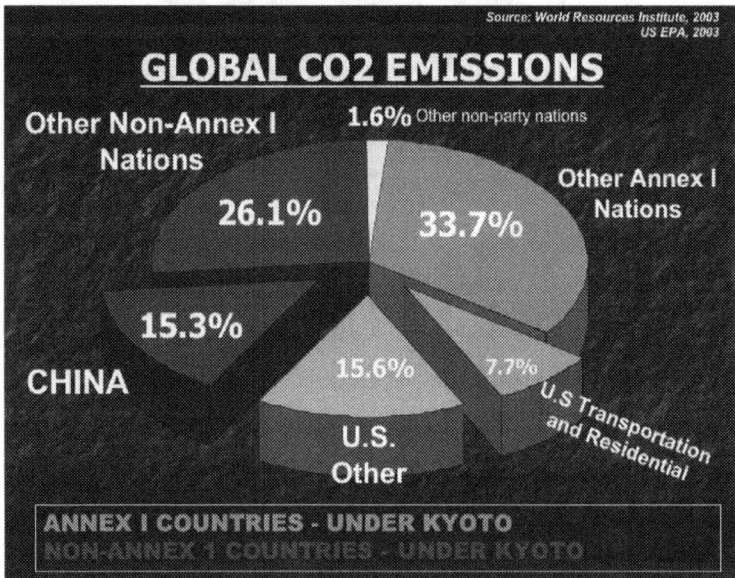

We must promote the green industry through tax incentives and investment in research and development.

Clean and renewable technologies, such as nuclear, wind, solar, and recycling provide ways to address climate change through economic stimulus.

Visiting countries such as Namibia reveals the dangers posed by climate change. Namibia is ranked among countries likely to be the most vulnerable to climate change and is expected to face a water scarcity by 2020.

The U.S. should take an active roll in developing
countries like Namibia to assist them develop in an
environmentally responsible manner.

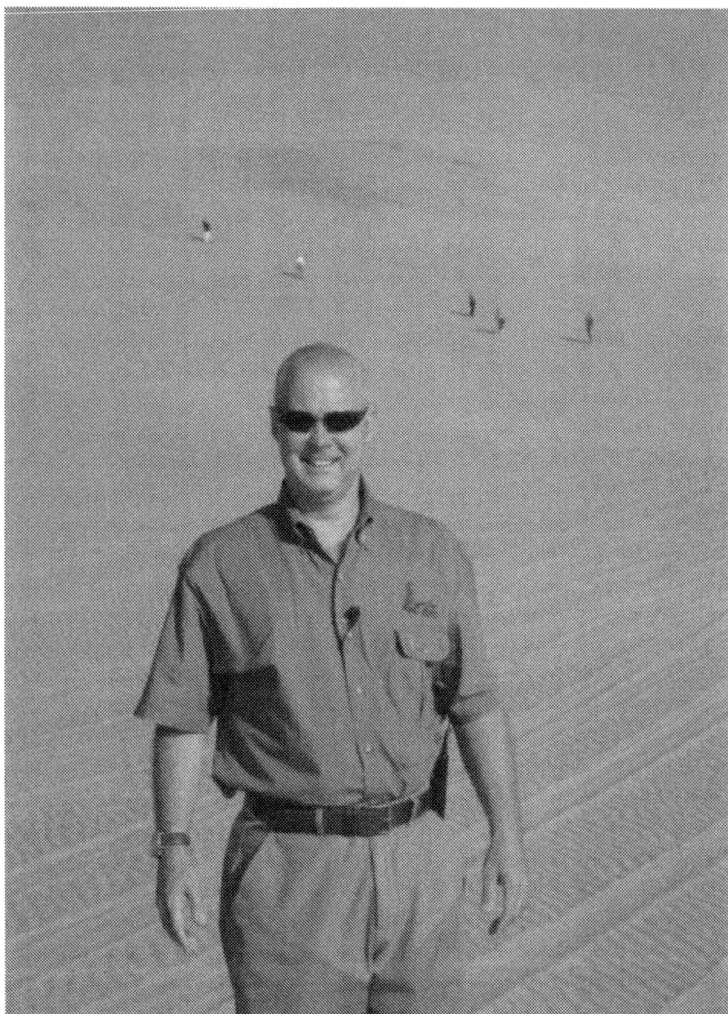

Endnotes

[1] Speth, James Gustave. *Red Sky at Morning: American and the Crisis of the Global Environment* (Yale University Press, 2004), xi-xii.

[2] Nosich, Gerald M. *Learning to Think Things Through: A Guide to Critical Thinking Across the Curriculum.* 2nd Ed. (Upper Saddle River, NJ: Pearson Prentice Hall, 2005), 22.

[3] "1992 Los Angeles Riots." (*Wilkiepedia*, 2006), <http://en.wikipedia.org/wiki/1992_Los_Angeles_riots> (accessed June 15, 2006).

[4] Mullikin, Thomas and Nancy Sara Smith. "Community Participation in Environmental Protection." *UCLA Journal of Environmental Law and Policy.* 75 (2002): 81-95.

[5] *Id.*

[6] Mullikin, Thomas Stowe, Nancy Sara Smith, and Michael Thomas Champion. "Inextricably Intertwined: Environmental Management and the Public." *Georgetown International Environmental Law Review* (Spring 2005): 1-44. <http://www.findarticles.com/p/articles/mi_qa3970/is_200504/ai_n13643932/pg_44> (accessed July 16, 2006).

[7] *Id.*

[8] *Id.*

[9] The Dalai Lama and Galen Rowell. "Universal Responsibility and the Environment." *The Government of Tibet in Exile.* 14 August 1996. <http://www.tibet.com/Eco/dleco2.html> (accessed June 16, 2006).

[10] *Rio Declaration on Environment and Development*, UN Conference on Environment and Development, 6th Sess., Annex I at A/CONF.151/26 (Vol. I) (12 August 1992).

[11] Basel Convention on the Control of Transboundary Movements of Hazardous Wastes and Their Disposal, Mar. 22, 1989, 1673 U.N.T.S. 125; 28 I.L.M. 657 (1989).

[12] Kyoto Protocol to the United Nations Framework Convention on Climate Change, Dec. 10, 1997, U.N. Doc FCCC/CP/1997/7/Add.1, 37 I.L.M. 22 (1998).

[13] Rotterdam Convention on the Prior Informed Consent Procedure for Certain Hazardous Chemicals and Pesticides in International Trade (10 September 1998).

[14] *UNECE Convention on Access to Information, Public Participation in Decision-making and Access to Justice in Environmental Matters,* Aarhus Convention (25 June 1998).

[15] *Id.*

[16] *Johannesburg Declaration on Sustainable Development,* World Summit on Sustainable Development in Johannesburg, South Africa (2-4 September 2002).

[17] *Rio Declaration,* principle 10.

[18] *Rio Declaration,* Agenda 21, section 23.2. (Also adopted at UNCED)

[19] Gore, Al. *Earth in the Balance, Ecology and the Human Spirit.* (Rodale Books, 2006), 354-55.

[20] *Id.* at 352.

[21] The TIME/ABC News/Stanford University poll was conducted by telephone March 9-14, 2006 among a random national sample of 1,002 adults. The results have a three-point error margin.

[22] Ledbetter, Jim. "Why Should I Be Good?" *TIME Europe Magazine.* June 5, 2006. <http://www.time.com/time/europe/magazine/article/0,13005,901060605-1198868,00.html> (accessed June 16, 2006).

[23] Gunther, Marc. "Global Warming Could Melt Your Portfolio." *CNNmoney.* March 21, 2006. <http://money.cnn.com/2006/03/21/news/international/pluggedin_fortune/> (accessed June 16, 2006).

[24] *Poll: Americans See a Climate Problem.* Posted Sunday, Mar. 26, 2006. <http://www.time.com/time/nation/article/0,8599,1176967,00.html> (accessed June 16, 2006).

[25] Browne, John. Stanford Business School, Stanford, CA. 19 May 1997. <http://www.gsb.stanford.edu/community/bmag/sbsm0997/feature_ranks.html> (accessed June 16, 2006).

[26] Michaels, Dr. Patrick J., *Meltdown: The Predictable Distortion of Global Warming by Scientists, Politicians, and the Media,* 214.

[27] *Climate Change: Global Problems, Global Solutions.* Prod. Thomas S. Mullikin. DVD. Moore and Van Allen, 2006.

[28] Kluger, Jeffrey. "The climate is crashing, and global warming is to blame. Why the crisis hit so soon--and what we can do about it." *TIME Magazine.* March 26, 2006.

[29] Kolbert, Elizabeth. "The Climate of Man—I." *The New Yorker.* April 2, 2005:56-71, 58.

[30] Gunther, "Global Warming Could Melt Your Portfolio."

[31] The Center for Health and the Global Environment at Harvard Medical School, Swiss Re and the United Nations Development Programme. *Climate Change Futures: Health, Ecological and Economic Dimensions.* November 2005. <http://www.climatechangefutures.org/> (accessed June 16, 2006).

[32] Carey, John and Sarah R. Shapiro. "Global Warming," *Businessweek.* August 16, 2004. <http://www.businessweek.com/magazine/content/04_33/b3896001_mz001.htm> (accessed June 16, 2006).

[33] Miller, Kevin. "Global warming's effects on Maine examined; Portland conference urges planning, technology." *Bangor Daily News (Maine),* May 11, 2006. <http://www.innovations.harvard.edu/news/12390.html> (accessed June 16, 2006).

[34] *Id.*

[35] Carey and Shapiro, "Global Warming."

[36] Natural Resources Defense Council. "Northeast States Lead Regional Global Warming Initiative." March 28, 2005. <http://www.nrdc.org/media/pressreleases/050328.asp> (accessed May 31, 2006).

[37] "West Coast States Strengthen Joint Climate Protection Strategy." *Oregon.* November 18, 2004. May 31, 2006. <http://www.oregon.gov/ENERGY/GBLWRM/Regional_Intro.shtml> (accessed May 31, 2006).

[38] Holliday, Charles O., Jr. "Sustainable Growth 2004 Progress Report." *DuPont.* <http://www2.dupont.com/Social_Commitment/en_US/SHE/usa/us1.html> (accessed June 16, 2006).

[39] Whirlpool Corporation. *Pew Center: Global Climate Change.* June 5, 2006. <http://www.pewclimate.org/companies_leading_the_way_belc/company_profiles/whirlpool/> (accessed June 16, 2006).

[40] Little, Amanda Griscom. "It was Just My Ecomagination: GE Kicks Off Ambitious Green Initiative." *Muckraker: the Dirt on Environmental*

Politics and Policy. May 10, 2005.
<http://www.grist.org/news/muck/2005/05/10/little-ge/> (accessed
June 16, 2006).

[41] Carey and Shapiro, "Global Warming."

[42] Steven Ferrey, *Sustainable Energy, Environmental Policy, and
States' Rights: Discerning the Energy Future Through the Eye of the
Dormant Commerce Clause*, 12 N.Y.U. ENVTL. L.J., 529 (2004).

[43] *Id.* at Appendix (discussing targeted technologies in the states).

[44] *See generally* Ole Langniss and Ryan Wiser, "The renewables portfolio standard in Texas: an early assessment", Energy Policy 31 (2003).

[45] Steven Ferry, *Power Future*, 15 Duke Envtl. L. & Pol'y F. 285

[46] *Id.*

[47] In late 2000, the price of crude oil spiked $9 a barrel from January of
that same year, to a record $31.50 a barrel. By July 2006, the price of a
barrel of crude oil had risen to $70 a barrel—a $15 increase from January of 2006. *See* U.S. Energy Information Admin., "United States Spot
Price FOB Weighted by Estimated Import Volume (Dollars per Barrel),"
Petroleum Navigator,
http://tonto.eia.doe.gov/dnav/pet/hist/wtotusaw.htm (last visited Aug.
15, 2006).

[48] *See* Tom Petri, U.S. Representative for Wisconsin's 6th Congressional
District, "Energy," On the Issues,
http://www.house.gov/petri/issues/energy.htm (last visited Aug. 15,
2006)("A large number of factors are contributing to the record prices.
The increased world demand for crude oil from countries such as China
and India, as well as political instability in major oil producing regions
such as Iraq, Iran, Nigeria and Venezuela have all contributed to tight
international markets. Domestically, U.S. refining capability has yet to
recover from the hurricanes . . . [i]t clear that it is time we starting investing in our own domestic production infrastructure, as well as look
for new alternative energy solutions."); U.S.-Canada Power System Outage Task Force, Final Report on the August 14, 2003 Blackout in the
United States and Canada: Causes and Recommendations (Apr. 2004),
available at: https://reports.energy.gov/BlackoutFinal-Web.pdf.

[49] Fred Sissine, Resources, Science, and Industry Division, Congressional Research Service, Issue Brief No. IB10041, Renewable Energy:

Tax Credit, Budget, and Electricity Restructuring Issues, Summary (Nov. 29, 2001), *available at* http://ncseonline.org.

[50] *Id.*

[51] *Id.*

[52] Hon. Joe Barton, Speech in the House of Representatives (Dec. 6, 2006).

[53] Leachate is formed as water seeps through a landfill, mixing with materials in the soil to develop an undesirable or even toxic substance that may seep into groundwater. Cornwell Waste Management Institute, "Landfill Leachate," Trash Goes to School (1991), http://cwmi.css.cornell.edu/TrashGoesToSchool/Landfill.html.

[54] North Carolina Division of Pollution Prevention and Environmental Assistance, Costs and Benefits of Recycling for North Carolina, No. DPPEA-97-51 (Sept. 1997), *available at* http://www.p2pays.org/ref/01/00363.pdf.

[55] Tex. Util. Code, Section 39.904(d) (2006).

[56] Landfills are a low priority-land use, wasting land that may have a higher use for agriculture, housing, industry, or recreation. Cooperative Extension Service, Kansas State University, "Resources for Recycling," (1992)http://www.oznet.ksu.edu/library/agec2/C724.pdf.

[57] *See* U.S. EPA, No. EPA-430-N-02-002, Inside the Greenhouse: A State and Local Resource on Global Warming (Spring 2002), *available at* http://www.epa.gov/globalwarming/greenhouse/greenhouse17/wasteline.html; *see also* Delaware Dept. of Natural Res. & Envtl. Control, Division of Air & Waste Management, Benefits of Recycling Components of Household Trash (2004), *available at* http://www.dnrec.state.de.us.

[58] Delaware Dept. of Natural Res. & Envtl. Control, *supra* note 34.

[59] Pennsylvania Dept. of Envtl. Protection, Recycling Saves our Environment, http://www.dep.state.pa.us/dep/deputate/airwaste/wm/recycle/FACTS/benefits4.htm (last visited Aug. 15, 2006).

[60] *Id.*

[61] *Id.*

[62] *Id.*

[63] *Id.*

[64] National Recycling Coalition, "Recycling and Reuse Add Value to the U.S. Economy," Results of the National REI Study (July 2001), http://www.nrc-recycle.org/resources/rei/studyresults.htm.

[65] *Id.*

[66] National Recycling Coalition, U.S. Recycling Economic Information Study, at ES-4 (July 2001), *available at* http://www.nrc-recycle.org/resources/rei/docs/fullreireport.pdf.

[67] Bureau of Labor Statistics Current Employment Statistics Survey (June 2006), *available at* http://data.bls.gov.

[68] Bureau of Economic Analysis, U.S. Department of Commerce report "Personal income by major source and earnings by industry -- United States 2004."

[69] NRC, "Recycling and Reuse Add Value to the U.S. Economy," *supra* note 43.

[70] U.S. EPA, "Recycling Facts and Figures," Municipal Solid Waste, http://www.epa.gov/msw/recycle.htm#Figures (last visited Aug. 15, 2006).

[71] United States Environmental Protection Agency, "EPA's Position on the Energy Crisis," (January 1974) *available at* http://www.epa.gov/history/topics/energy/01.htm.

[72] S. 3654, 109th Cong. §2 (2006).

[73] *See generally* University of Michigan, Center for Sustainable Materials, "U.S. Material Use," *available at* http://css.snre.umich.edu.

[74] Environmental Protection Agency, Municipal Solid Waste, Steel, *available at* http://www.epa.gov/msw/steel.htm (Steel is a versatile commodity that plays a major part in everyday life—it is used in applications ranging from food cans and household containers to automobiles and office buildings. Steel makes up the largest category of metals in the municipal solid waste (MSW) stream.).

[75] Carey and Shapiro, "Global Warming."

[76] Linden, Eugene. "Global Warming Slips on Its Ski Mask," *Los Angeles Times*. July 30, 2006 <http://www.latimes.com/news/printedition/opinion/la-oe-linden30jul30,1,3547323.story?ctrack=1&cset=true> (accessed June 16, 2006).

[77] Carey and Shapiro, "Global Warming."

[78] Carey and Shapiro, "Global Warming."

[79] Little, "It was Just My Ecomagination."

[80] James B. Holderman, "The Need for an International Perspective" (Plenary address to the Council on International Programs, Chapel Hill, University of North Carolina, February 25, 1983).

[81] *Climate Change: Global Problems, Global Solutions*. Prod. Thomas S. Mullikin. DVD. Moore and Van Allen, 2006.

[82] Quoted in Adams J.M. Global land environments since the last interglacial. (Oak Ridge National Laboratory, TN, 1997). <http://www.esd.ornl.gov/projects/qen/nerc130k.html> (accessed August 1, 2006).

[83] Kluger, "The climate is crashing."

[84] Arctic Climate Impact Assessment (ACIA), *Impacts of a Warming Arctic*. Cambridge: Cambridge University Press, 2004.

[85] Speth, James Gustave. "Climate Change After the Elections: What We Can Do in America, A Ten Point Plan," *Red Sky at Morning: America and the Crisis of the Global Environment*. Boston: Yale University Press, 2005. 6.

[86] Shukman, David. "Greenland ice-melt 'speeding up'" *BBC News*. Wednesday, 28 July, 2004 <http://news.bbc.co.uk/1/hi/world/europe/3922579.stm> (accessed August 1, 2006).

[87] Holtz, Robert Lee. "Greenland's slip-sliding glaciers offer chilling evidence of warming," MercuryNews.com. June 25, 2006 <http://www.mercurynews.com/mld/mercurynews/news/world/14899 428.htm> (accessed August 1, 2006).

[88] Image created by Jonathan Overpeck and Jeremy Weiss, courtesy University of Arizona Department of Geosciences Environmental Studies Laboratory. *National Geographic News*. April 26, 2004 <http://news.nationalgeographic.com/news/2004/04/0420_040420_ earthday.html> (accessed August 1, 2006).

[89] Speth, "Climate Change After the Elections," 6.

[90] "Consequences of Global Warming," National Resources Defense Council. January 9, 2006. <http://www.nrdc.org/globalWarming/fcons.asp> (accessed August 1, 2006).

[91] Kluger, "The climate is crashing."

[92] *Id.*

[93] Eilperin, Juliet. "U.S. Wants Polar Bear Listed As Threatened," *The Washington Post*. December 27, 2006, A01.

[94] Westerling, Anthony L. and Dan Cayan, Scripps Institution of Oceanography, 2006.

[95] Moreira, Naila. "Study Links Increase in Wildfires to Global Warming," *The Boston Globe*. July 7, 2006. <http://www.boston.com/news/nation/articles/2006/07/07/study_lin ks_increase_in_wildfires_to_global_warming/> (accessed August 1, 2006).

[96] *Id.*

[97] *Impacts of Ocean Acidification on Coral Reefs and Other Marine Calcifiers*. July 5, 2006.

[98] Quoted in O'Hanlon, Larry. "Rising Ocean Acidity Threatens Reefs," *Discovery News*. July 5, 2006 <http://dsc.discovery.com/news/2006/07/05/acidocean_pla.html?cat egory=animals&guid=2006070517000> (accessed August 1, 2006).

[99] Editorial, "Multifaceted Threat," *The Houston Chronicle*. July 9, 2006 <http://www.chron.com/disp/story.mpl/editorial/4035105.html> (accessed August 1, 2006).

[100] Sun Tzu *The Art of War*

[101] *Climate Change: Global Problems, Global Solutions*. Prod. Thomas S. Mullikin. DVD. Moore and Van Allen, 2006.

[102] Adler, Jerry. "Going Green," *Newsweek*. July 17, 2006. 43.

[103] Speth, "Climate Change After the Elections," 11-12.

[104] Member countries include the EU 15 (Austria, Belgium, Denmark, Finland, France, Germany, Greece, Ireland, Italy, Luxembourg, the Netherlands, Portugal, Spain, Sweden and the United Kingdom) plus the Czech Republic, Hungary, Iceland, Norway, Poland, Slovak Republic, Switzerland and Turkey.

[105] French, Howard W. "Anger in China Rises over Threat to Environment," *The New York Times*. Tuesday 19 July 2005. <http://www.truthout.org/cgi-bin/artman/exec/view.cgi/34/12742/printer> (accessed August 1, 2006).

[106] Quoted in "The polluter pays: how environmental disaster is straining China's social fabric," *Financial Times of London*. 27 January 2006.

[107] *Id.*

[108] Scott, Robert. "U.S.-China Trade, 1989-2003: Impact on Jobs and Industries, Nationally and State-by-State." *Economic Policy Institute*. January 11, 2005. <http://usinfo.state.gov/ei/Archive/2005/Jan/12-31762.html> (accessed August 1, 2006).

[109] *Id.*

[110] *Id.*

[111] *International Monetary Fund (IMF) Articles of Agreement.* Adopted at the United Nations Monetary and Financial Conference, Bretton Woods, New Hampshire, July 22, 1944. Entered into force December 27, 1945. Amended effective July 28, 1969, by the modifications approved by the Board of Governors in Resolution No. 23-5, adopted May 31, 1968; amended effective April 1, 1978, by the modifications approved by the Board of Governors in Resolution No. 31-4, adopted April 30, 1976; and amended effective November 11, 1992, by the modifications approved by the Board of Governors in Resolution No. 45-3, adopted June 28, 1990. <http://www.imf.org/external/pubs/ft/aa/index.htm> (accessed August 1, 2006).

[112] *Id.*

[113] Roberts, Paul Craig. "Issues & Insights: Globalism, End Of Socialism Causes Of Jobless Recovery." *Investors Business Daily.* Friday, August 30, 2002 <http://www.vdare.com/roberts/globalism.htm> (accessed August 1, 2006).

[114] *Id.*

[115] Roberts, Paul Craig. "A Nation of Waitresses and Bartenders?" *The Progress Report.* <http://www.progress.org/2006/econo2.htm> (accessed August 1, 2006).

[116] *Climate Change: Global Problems, Global Solutions.* Prod. Thomas S. Mullikin. DVD. Moore and Van Allen, 2006.

[117] *Id.*

[118] "India: Environmental Issues," Energy Information Administration. February 2004 <http://www.eia.doe.gov/emeu/cabs/indiaenv.html> (accessed August 1, 2006).

[119] *Id.*

[120] Speth, James Gustave. *Red Sky at Morning: American and the Crisis of the Global Environment* (Yale University Press, 2004), xi-xii.

Works Cited

"1992 Los Angeles Riots." *Wilkiepedia* (2006). http://en.wikipedia.org/wiki/1992_Los_Angeles_riots (accessed June 16, 2006).

Adler, Jerry. "Going Green," *Newsweek*. July 17, 2006. 43.

Arctic Climate Impact Assessment (ACIA), *Impacts of a Warming Arctic*. Cambridge: Cambridge University Press, 2004.

Carey, John and Sarah R. Shapiro. *Businessweek*. August 16, 2004. http://www.businessweek.com/magazine/content/04_33/b3896001_mz001.htm

The Center for Health and the Global Environment at Harvard Medical School, Swiss Re and the United Nations Development Programme. *Climate Change Futures: Health, Ecological and Economic Dimensions*. http://www.climatechangefutures.org/

Climate Change: Global Problems, Global Solutions. Prod. Thomas S. Mullikin. DVD. Moore and Van Allen, 2006.

"Consequences of Global Warming," National Resources Defense Council. January 9, 2006. http://www.nrdc.org/globalWarming/fcons.asp

The Dalai Lama and Galen Rowell. "Universal Responsibility and the Environment." *The Government of Tibet in Exile*. 14 August 1996. http://www.tibet.com/Eco/dleco2.html

Editorial, "Multifaceted Threat," *The Houston Chronicle*. July 9, 2006 http://www.chron.com/disp/story.mpl/editorial/4035105.html

French, Howard W. "Anger in China Rises over Threat to Environment," *The New York Times*. Tuesday 19 July 2005.

http://www.truthout.org/cgi-bin/artman/exec/view.cgi/34/12742/prin
ter

Gore, Al. *Earth in the Balance: Ecology and the Human Spirit.* Rodale
Books, 2006.

Gunther, Marc. "Global Warming Could Melt Your Portfolio."
CNNmoney. March 21, 2006.
http://money.cnn.com/2006/03/21/news/international/pluggedin_for
tune/

Holliday, Charles O., Jr. "Sustainable Growth 2004 Progress Report."
DuPont.
http://www2.dupont.com/Social_Commitment/en_US/SHE/usa/us1.h
tml

Holtz, Robert Lee. "Greenland's slip-sliding glaciers offer chilling
evidence of warming," MercuryNews.com. June 25, 2006.
http://www.mercurynews.com/mld/mercurynews/news/world/148994
28.htm

"India: Environmental Issues," Energy Information Administration.
February 2004.
http://www.eia.doe.gov/emeu/cabs/indiaenv.html

Kluger, Jeffrey. "The climate is crashing, and global warming is to
blame. Why the crisis hit so soon--and what we can do about it." *TIME
Magazine.* March 26, 2006.

Kolbert, Elizabeth. "The Climate of Man—I." *The New Yorker.* April 2,
2005:56-71.

Ledbetter, Jim. "Why Should I Be Good?" *TIME Europe Magazine.*
June 5, 2006.
http://www.time.com/time/europe/magazine/article/0,13005,9010606
05-1198868,00.html

Little, Amanda Griscom. "It was Just My Ecomagination: GE Kicks Off Ambitious Green Initiative." *Muckraker: the Dirt on Environmental Politics and Policy.* May 10, 2005.

Miller, Kevin. "Global warming's effects on Maine examined; Portland conference urges planning, technology." *Bangor Daily News (Maine),* May 11, 2006.
http://www.innovations.harvard.edu/news/12390.html

Moreira, Naila. "Study Links Increase in Wildfires to Global Warming," *The Boston Globe.* July 7, 2006.
http://www.boston.com/news/nation/articles/2006/07/07/study_link s_increase_in_wildfires_to_global_warming/

Mullikin, Thomas and Nancy Sara Smith. "Community Participation in Environmental Protection." *UCLA Journal of Environmental Law and Policy.* 75 (2002): 81-95.

Mullikin, Thomas Stowe, Nancy Sara Smith, and Michael Thomas Champion. "Inextricably Intertwined: Environ-mental Management and the Public." *Georgetown Inter-national Environmental Law Review* (Spring 2005): 1-44.

Natural Resources Defense Council. "Northeast States Lead Regional Global Warming Initiative." March 28, 2005. May 31, 2006.
http://www.nrdc.org/media/pressreleases/050328.asp

Nosich, Gerald M. *Learning to Think Things Through: A Guide to Critical Thinking Across the Curriculum.* 2nd Ed. Upper Saddle River, NJ: Pearson Prentice Hall, 2005.

O'Hanlon, Larry. "Rising Ocean Acidity Threatens Reefs," *Discovery News.* July 5, 2006.
http://dsc.discovery.com/news/2006/07/05/acidocean_pla.html?categ ory=animals&guid=20060705170030

Rio Declaration on Environment and Development, UN Conference on Environment and Development, 6th Sess., Annex I at A/CONF.151/26 (Vol. I) (12 August 1992).

Roberts, Paul Craig. "Issues & Insights: Globalism, End Of Socialism Causes Of Jobless Recovery." *Investors Business Daily*. Friday, August 30, 2002.
http://www.vdare.com/roberts/globalism.htm

Roberts, Paul Craig. "A Nation of Waitresses and Bartenders?" *The Progress Report*.
http://www.progress.org/2006/econo2.htm

Scott, Robert. "U.S.-China Trade, 1989-2003: Impact on Jobs and Industries, Nationally and State-by-State." *Economic Policy Institute*. January 11, 2005.
http://usinfo.state.gov/ei/Archive/2005/Jan/12-31762.html

Shukman, David. "Greenland ice-melt 'speeding up'" *BBC News*. Wednesday, 28 July, 2004.
http://news.bbc.co.uk/1/hi/world/europe/3922579.stm

Speth, James Gustave. "Climate Change After the Elections: What We Can Do in America, A Ten Point Plan," *Red Sky at Morning: America and the Crisis of the Global Environment*. Boston: Yale University Press, 2005.

"West Coast States Strengthen Joint Climate Protection Strategy." *Oregon*. November 18, 2004. May 31, 2006.
http://www.oregon.gov/ENERGY/GBLWRM/Regional_Intro.shtml

Whirlpool Corporation. *Pew Center: Global Climate Change*. June 5, 2006.
http://www.pewclimate.org/companies_leading_the_way_belc/company_profiles/whirlpool/

Additional Suggested Reading

Catton, Jr., William R. *Overshoot: The Ecological Basis of Revolutionary Change*. Urbana: University of Illinois Press, 1980.

Durning, Alan. *How Much is Enough?: The Consumer Society and the Future of the Earth*. New York: W. W. Norton, 1992.

Elvin, Mark. *The Retreat of the Elephants: An Environmental History of China*. New Haven: Yale University Press, 2004.

Gore, Albert. *Earth in the Balance: Ecology and the Human Spirit*. Boston: Houghton Mifflin, 1992.

Graham, John D. and Jennifer Kassalow Hartwell. *The Greening of Industry: A Risk Management Approach*. Cambridge, Mass.: Harvard University Press, 1997.

Houghton, John. *Global Warming: The Complete Briefing*, third edition. Cambridge: Cambridge University Press, 2004.

Leggett, J. *The Carbon War: Global Warming and the End of The Oil Era*. New York: Routledge, 2001.

Shutkin, William A. *The Land That Could Be: Environmentalism and Democracy in the Twenty-First Century*. Cambridge, Mass.: MIT Press, 2000.

Speth, James Gustave. *Red Sky at Morning: America and the Crisis of the Global Environment*. Boston: Yale University Press, 2005.

Switzer, Jacqueline Vaughn. *Green Backlash: The History and Politics of Environmental Opposition in the U.S.* Boulder, Col.: Lynne Rienner Publishers, 1997.

Weart, Spencer R. *The Discovery of Global Warming*. Cambridge, Mass.: Harvard University Press, 2003.

Appendix A

Inextricably Intertwined – Environmental Management and the Public

Tom Mullikin*, Nancy Smith**, Michael Champion***

The Georgetown International Environmental Law Review Volume 17, Issue 3, Pages 393-433 (Spring 2005).

Reprinted with Permission of the Publisher, Georgetown International Environmental Law Review © 2005.

* J.D., 1986, University of South Carolina School of Law; B.A., 1982 cum laude, Phi Beta Kappa, University of South Carolina. Mr. Mullikin is a member of Moore & Van Allen PLLC and practices in the firm's Charlotte office. As leader of the firm's Government, Policy and Regulatory Affairs group, his practice focuses on the intersection among governmental, environmental, and business interests.

** J.D., 1998, University of North Carolina at Chapel Hill School of Law, Member of the Holderness Moot Court Negotiations Team; B.A., 1995, University of North Carolina at Chapel Hill. Ms. Smith is an associate of Moore & Van Allen PLLC in the Charlotte office where she practices with the firm's Government, Policy and Regulatory Affairs group.

*** J.D., 2004, University of North Carolina at Chapel Hill School of Law, Executive Editor of the North Carolina Journal of International Law and Commercial Regulation; M.A., 1999 summa cum laude, University of North Carolina at Charlotte; B.A., 1996 magna cum laude, University of North Carolina at Charlotte. Mr. Champion is an associate of Moore & Van Allen PLLC in the Charlotte office practicing with the firm's environmental practice group.

The authors would like to express their deep gratitude towards Natasha Beam and Whitney Robinson for their hard work and assistance in the research and preparation of this article.

I. INTRODUCTION

Over the last forty years, the idea that the public holds a legal right to participate generally in government decision-making has gained wide acceptance both domestically and internationally. 1 This fact is especially true within the realm of environmental law, where government decisions greatly impact the environment in which all citizens must live. Generally, pro-participation arguments presuppose that because every non-natural act that affects the environment involves government decision-making at some level, the views of the public citizenry inhabiting that environment and consequently bearing the cost of its degradation should be considered during development and implementation of a workable policy. 2

[*395] The growing acceptance of environmental public participation rights is evident in the arena of international law 3 where recent international environmental initiatives have formed a critical mass around the evolving concept of sustainable development. 4 These broad initiatives, and the large number of specific legal instruments they spawn, place enormous import on the concept of environmental public participation as a key ingredient of sustainable development. 5 The growing international discourse has influenced many international aid agencies and private institutional investors to view the concept of sustainable development as the proper means towards achieving development in transitional States. For example, the United States Agency for International Development (USAID), the World Bank, and the Asian Development Bank have all formulated policies that require aid recipients and developers to adhere to certain principles and guiding rules promoting sustainable development as the concept has been recognized and codified by international law. 6 Consequently, these organizations and others mandate public participation as a key factor of sustainable development, and their funding is often predicated, at least in part, on how well developers and their host nations involve the public in project decision-making.

Many States, both developed and transitional, have also enacted broad statutory provisions providing for environmental public participation rights. Some have even adopted constitutional amendments to guarantee environmental public participation rights in governmental decision-making. 7 For example, public participation is a cornerstone concept found throughout U.S. environmental law. The Environmental Protection Agency (EPA) has created several non-regulatory programs aimed at including the public in meaningful ways in the decision-making process. 8 Likewise, many States in the European Union and the European Community have begun to implement ambitious public participation provisions of the United Nations Economic Commission for Europe (UNECE) Convention [*396] on Access to Information (Aarhus Convention). 9 One may also find broad environmental public participation rights in the developing world, where transitional States in Latin America, Asia, and the Middle East are working to develop environmental regulatory schemes guaranteeing public involvement in governmental decision-making. 10

While growing more popular, the concept of environmental public participation is not universally applied. Although many political systems purport to make some provision for environmental public participation, the scope and effect of that participation is inconsistent. 11 The contours of a public "right" to participate in environmental decision-making remain dynamic and uncertain.

This paper advances a basic argument: environmental public participation rights are necessary to industrial development, and deliberate, effective, and well-reasoned economic growth can occur only after involvement of critical host-community stakeholders. Such rights should not be ignored by the governments of transitional States, nor should they be ignored by industry seeking to invest in those transitional States. Three factors support this position. First, and perhaps most importantly, there is a crucial nexus between the industrialization of society and the need for environmental regulation containing assurances of public participation in environmental decision-making. 12 Where industrial development occurs, public apprehension and concern for the protection of air, land, and water eventually follow. As States move through the stages of industrialization, the resulting environmental stressors intensify concern for the protection of the environment among a public that ultimately bears many of the externalities of development. Failure

A-3

to recognize this phenomenon and to include the public as participants in environmental decision-making will have long-term negative consequences on industry. Therefore, public support is crucial to continued and sustainable development.

This "nexus" between public participation and successful economic growth is reflected in the historical experiences of developed and heavily industrialized States such as the United States and the States that make up the European Union. It is no coincidence that these States have created elaborate mechanisms for environmental public participation rights and have been the most successful at injecting those rights into broad, far-reaching environmental legislation. 13 A [*397] study of this historical nexus demonstrates how public concern resulting from industrialization created the impetus for the current public participation schemes embedded in the environmental legislation of these developed nations.

Second, both international and internal forces are compelling transitional States to secure environmental participatory rights for their citizens. International support for environmental participatory rights has resulted from the recognition that environmental public participation rights are a key factor for sustainable development. 14 Many transitional States have bound themselves by treaty to include the public in governmental decisions regarding the environment. Further, transitional States are finding it increasingly difficult to secure funding and aid for development without adhering to at least some minimum international standard of public involvement in decisions concerning development projects. In addition to international pressure, transitional States increasingly face internal pressure to implement environmental public participation rights as a means to mitigate the potentially disastrous consequences of public discord (e.g., rioting). 15 When the affected public is left out of the development process, transitional governments face potentially devastating consequences and long-term costs. The result is that transitional States have begun to recognize the importance of public involvement and are responding accordingly.

Finally, transitional States and private industry should not ignore public participation rights because the creation of these rights attracts capital investors. Capital investors generally recognize that the public's ability to participate in environmental decision-making produces positive economic outcomes essential for successful development. 16 Environmental public participation rights make good economic sense. Markets require a free flow of information. Investors like stability and seek proper risk allocation for their large capital outlays. When States implement public participation schemes, they facilitate information flow and ensure stability by creating a rule of law. As discussed above, the public's ability to participate can also alleviate public discord and promote acceptance of large infrastructure projects, another key factor for market stability.

This paper will examine these three factors in order to demonstrate the importance of environmental public participation rights to well-reasoned and sound development. Part II demonstrates how the historical progression of industrialization produced environmental stress that created the impetus for environmental regulation and the public's right to participate in that regulation. While this section focuses on Western examples, the nexus between industrialization and environmental public participation rights is universal. Therefore, the analysis is helpful to understand how environmental regulation emerges generally in States marching towards development.

[*398] Part III examines the product of this historical nexus by surveying actual public participatory rights found in the major pieces of EU and U.S. environmental legislation. This section argues that environmental public participation rights play a crucial role in developed States, and may serve as effective templates for transitional States seeking to implement such rights.

Part IV traces the international emergence of environmental public participation rights as a factor for sustainable development. As international law increasingly reflects a developed and refined concept of sustainable development, it builds a foundation for public participation rights in environmental decision-making. This section demonstrates how the interrelated concepts of sustainability and participation have deeply influenced binding regional agreements and international aid policies for development. Consequently, transitional States, bound by the formal mandates of these agreements and aid policies, are implementing environmental public participation rights to an unprecedented degree. This section also demonstrates how overwhelming internal pressure from concerned citizens forces transitional States to involve the public in environmental decision-making.

The final section explains the importance of enacting environmental public participation rights as a means of mitigating risk and allocating the cost of development among several stakeholders. The section concludes that industrial investors who build capacity in transitional States should recognize and encourage the implementation of such rights to ensure sustainable development.

II. THE HISTORICAL NEXUS BETWEEN INDUSTRIAL DEVELOPMENT, ENVIRONMENTAL REGULATION, AND PUBLIC PARTICIPATION

Strong public concern for the environment, as a resource requiring protection from unmitigated human activity, is a relatively new development which has appeared only within the last forty years. Even more recent is the notion that the public should play an integral role in developing and enforcing an environmental regulatory scheme designed to alleviate environmental degradation, a notion that has yet to take root in many areas of the world. 17 To determine why the concept of environmental public participation has gained strong support in some countries, it is useful to trace the broader historical forces that have played a role in the development of public demand for environmental accountability. History shows that mounting public concern and demand for public participation rights result from the rise of environmental stress, 18 stemming largely from under-regulated [*399] industrial growth, and greater scientific understanding of the effects of environmental stressors.

Environmental regulation 19 arises from concern about a public problem. This situation inherently requires three conditions: (1) the existence of one or more environmental stressors that affect the environment through nuisance, degradation, or other means; (2) the awareness (through scientific knowledge, education, ready dissemination of information, etc.) that environmental stressors are the root cause of environmental degradation; and (3) the existence of sufficient concern to remedy that degradation. 20 Such a combination of factors came together as a result of industrial revolutions, and the trend continues throughout the developed world with the rise of globalization.

While a survey of the events that led to the necessary combination of factors could form a multi-volume treatise, what follows is a simplification of these events, provided to ground recent environmental regulation and public participation rights in a deeper historical tradition. Using the industrialized West as an example, this section outlines the historical nexus between environmental stress caused by industrial development and calls by a concerned public for greater participation in environmental decision-making. While this framework is modeled after Western industrialization, it is also useful for understanding the ways in which the environmental effects of industrialization generally lead to calls for greater public participation rights and can be applied to current transitional States with a few modifications.

A. STAGE ONE: DEVELOPING ENVIRONMENTAL CONSCIOUSNESS

Environmental concern requires consciousness of environmental problems. In the Western tradition, environmental consciousness developed gradually over centuries. 21 Ancient cultures throughout the world were at least cognizant of the problems associated with waste management and resource conservation. 22 In fact, many of the Greek city-states adopted waste management ordinances that required citizens to dispose of their trash outside their cities. 23 Both the Greeks [*400] and Romans built extensive and costly sewage systems that likely required numerous decisions at the governmental level regarding policy implementation. 24 Evidence even suggests that there was some level of public participation throughout this process, if only through the normal democratic processes of the civitas. 25

However, Greeks and Romans were concerned more about dealing with the nuisance of human and animal waste than they were with the environmental effects of such waste. The chain of cause and effect between development, environmental degradation, and its negative impact on human health was not clearly understood. Humans did not consider the environment because they were not conscious of any real threat to the environment. Rather, the sewage systems were a common sense response to alleviate the unpleasant effects of waste by disposing of it responsibly.

The lack of any significant environmental consciousness continued throughout the Early and High Middle Ages. As the feudal hierarchical structure replaced the concept of the civitas and Christian anthropocentrism replaced scientific inquiry, political and social discourse involving public participation generally, and the protection of the environment specifically, were a dead letter. 26 Medieval Europeans not only lacked the resources and scientific knowledge to appreciate the relationship between environmental degradation and human health, but they also lacked any real political and social power to do anything about it.

However, two key historical events of the Middle Ages spurred the slow but consistent development of Western consciousness regarding the environment and the human impact upon it: the rise of the Medieval city and the evolution of the university. 27 The rise of the city increased trade and industry, created the framework for the modern market economy and produced environmental stress that impressed upon the public the notion that industry and overpopulation carried environmental consequences. The universities laid the foundation for the emergence of scientific inquiry and discovery and a more secular notion of critical analysis. Both movements would later become essential during the Renaissance and the so-called Scientific Revolution, and both were instrumental in the coming age of industrialization. 28 These events eventually ushered in a [*401] move away from Christian anthropocentrism and toward humanism-the idea that humans should and must intervene in earthly events that affect them. 29

1. Rise of the Medieval City

While towns and cities had formed an important function in Roman society, their numbers declined dramatically throughout the Middle Ages as society became increasingly fragmented and the economy moved towards agricultural subsistence. 30 By the period of the High Middle Ages, the development of new and more efficient agricultural practices led to an increase in food production and a general surplus in the food supply. Free from the need to produce their own food, many Europeans were eager to embark upon other economic functions and fled to emerging towns, often in the old Roman centers, where they could group together and pursue those functions. 31 These migrations effectively reinvigorated commerce and, as money began once again to circulate, craftspeople and merchants flourished. Their increasing numbers created newer, larger, and evolving markets, and the Medieval cities were born. 32

Additionally, the food surplus kept food prices low and allowed townspeople to spend larger portions of their income on goods, ensuring the continued growth of markets. The aristocracy, eager to seize an opportunity to gain additional tax revenue, pledged resources to protect trade routes between cities. 33 While cities and markets revived only gradually, a regular exchange of goods had fully developed among the major population centers of Europe by the fourteenth century. 34 In fact, historians often speak of this period as one of "commercial revolution" whereby the system of new commercial practices and institutions developing alongside the expanded volume of trade eventually transformed all Western society. 35 This "commercial revolution" essentially formed the building blocks for the modern market economy and the foundation for industrialization.

Perhaps the most important aspect of the growth of cities during the Middle Ages was the rise of a quasi-environmental consciousness among these cities' inhabitants. The physical environment of medieval cities was unpleasant. 36 Many of the early trades involved the slaughtering and tanning of animals. These industries, once they had become concentrated in cities, dumped vast amounts of [*402] blood, tannic acids, fat, hair, and other waste products into the rivers surrounding cities. 37 Overcrowding was also an issue. Medieval towns were not allowed to exceed their chartered boundaries, and citizens did not want to relocate far beyond their city's defensive walls. As populations grew, space became limited. 38 City residents lived in cramped and unsanitary conditions where human and animal wastes and garbage were freely deposited on the streets, causing disease and poor health. 39 Further, with the larger urban populations burning wood and coal fires, high concentrations of smoke and soot created harmful amounts of noxious fumes. 40

Townspeople became increasingly concerned and vocal about the growing pollution problems caused by overpopulation and burgeoning industrial development. 41 Fearing rebellion, or perhaps responding to it, city council members attempted to regulate industrial practices. 42 For example, some towns enacted ordinances prohibiting the burning of coal within one mile of the town boundary. 43 Others mandated that certain industries pollute the water downstream of the city. 44

These regulations became even more widespread after the Black Death reaped devastating consequences on Europeans, wiping out over one-half of the total population of medieval Europe. 45 One immediate product of the plague was an awakening among townspeople that greater regulation of urban activity was necessary for public health. 46 While most of these regulations centered on promoting cleanliness and sanitation among the citizens themselves, many carried over to the trades and industries on the premise that all citizens of a town, including industry, had a responsibility to organize their affairs in a sanitary manner to promote the public health. 47

[*403]

2. The Rise of the University

The emergence of the cities also instigated the rise of the university. 48 Unlike earlier monastic schools that existed throughout the Middle Ages, the universities, while church-based, were meant to be seats of learning for non-monastic clergy and laymen who sought education without wanting to become clergy. Perhaps the most notable aspect of the rise of the university is that it sparked a revival of interest in works of classical antiquity. 49 While this resurgence was not as deep or far-reaching as that which took place during the Renaissance, it was a significant step towards understanding and developing concepts of democracy and republicanism, key ingredients for the future development of public participation in government decision-making. Further, the universities generated a growing and intense interest in scientific discovery and the desire to understand the human condition through reason, an element crucial for later scientific inquiry into the effects of industrialization on environmental degradation.

Like the ancient Greeks and Romans before them, citizens of pre-industrial European society were still not concerned for the environment. Any clear understanding of how man-made environmental stressors impacted the environment and public health was limited to issues of sanitary improvement and resource management. However, the growth of medieval and renaissance cities and the contributions of burgeoning scientific thought contributed to an increasing concern in public health issues as well as a growing awareness regarding resource management and disease prevention.

As Europeans moved toward industrialization, the rise of the cities and universities precipitated both a growing environmental consciousness and a desire to stem degradation. These events also provided a catalyst for the Enlightenment and the Scientific Revolution, two movements that would play a seminal role in advancing scientific thought, technology, and republican idealism throughout the West. 50 These elements, both a new environmental consciousness and greater scientific understanding, were necessary preconditions for environmental regulation in Western society. 51 The exponential growth of environmental stress resulting from the Industrial Revolutions would eventually spark broad environmental concern and subsequent demand for environmental regulation.

B. STAGE TWO: INDUSTRIALIZATION, ENVIRONMENTAL STRESS, AND GROWING PUBLIC CONCERN OVER DEGRADATION

Between 1850 and 1900, industrialization had fully come of age in most [*404] European States and the United States. 52 Building on the earlier achievements and innovations of the Industrial Revolution in Britain, 53 Western States were experiencing an age of solid economic expansion and prosperity. 54 As fortunes grew, so too did domestic and foreign markets for goods and services. The growth of trade that began with the rise of the medieval city blossomed into a full-scale market economy based upon a concept of free trade and interlocking, self-perpetuating consumer markets. 55

Economic growth was supplemented by the existence of liberal free trade and laissez-faire policies pursued by recently centralized Western governments seeking to increase the tax base and promote a free flow of trade. 56 At the core of this pro-development, laissez-faire attitude among Western policy-makers was a belief that industrialization was the ultimate realization of human progress and that a faithful adherence to market forces, technological progress, and industrial development would cure all evils and build national prestige. 57

Policy-makers felt that large-scale unrestricted industrialization was the surest way to develop a nation's strength. In many ways the facts substantiated this belief. By 1900, for instance, citizens of Europe and the United States were the richest consumers in the world. 58 Dramatic increases in population did not result in a lack of resources; rather, a steady rise in national incomes mostly resulted in a surplus of food, goods, and services. 59 The national incomes of Britain, Germany, and the United States more than doubled between 1850 and 1900. 60 Real wages increased in Britain by sixty-six percent during that same period. 61 Because transportation and production costs were cheaper, regulation was low, and labor was plentiful, the overall costs of goods and services declined to a point where nearly all segments of the domestic markets could purchase them. 62

However, for all its progress, the nearly unregulated industrialization had serious environmental consequences for citizens of industrializing States and produced a number of environmental stressors providing the impetus for public concern and calls for greater environmental regulation.

[*405]
1. Overcrowding and Disease

By 1900, Western population had increased dramatically. In Europe, the population increased from 270 million to 460 million between 1850 and 1910. 63 Urban populations grew much faster than rural populations as people migrated to the cities out of economic necessity. 64 Not surprisingly, overcrowded conditions caused by dramatic increases in population created an unhealthy environment and poor quality of life for urban dwellers. While this was not a new phenomenon for rural Europeans, who for centuries endured poor living conditions, such rapid urbanization magnified the situation. 65 Dirty habits that might have been acceptable in the scattered countryside had disastrous and deadly consequences in densely packed tenement buildings. 66 Cities had neither the knowledge nor the capacity to deal with the problems of overcrowding, waste, and pollution associated with industrialization.

Surveys and Poor Commission reports from Britain during the middle of the nineteenth-century provide insight into the plight of the working-class urban dwellers in the larger industrial cities. A report on working-class housing in the city of Birmingham in 1843 describes the condition of working-class row houses in the city:

> [They] are extremely numerous; ... a very large
> number of the poorer classes of the inhabitants reside
> in them ... Most of these houses are three stories

high, and built, as it is termed, back to back. There is a wash-house, an ash-pit, and a privy at the end ... and not infrequently one or more pigsties and heaps of manure. Generally, speaking, the privies ... are in a most filthy condition. Many which we inspected were in a state which renders it impossible for us to conceive how they could be used; they were without doors and overflowing with filth. 67

Other Poor Commissions reported that many urban tenants were forced to live in cellars under the ground with little ventilation and no more than twelve to fifteen square feet of living space. 68 Often, observers noted that six to eight people would share a room no more than seventy square feet. 69

The lack of municipal control over sanitary measures and political will to implement environmental regulation perpetuated the unhealthy conditions and led to a host of deadly diseases. 70 Various Commission reports and primary [*406] accounts note that, without local regulation, city streets were regularly used as sewers and open drains. One observer noted that the streets were gutters onto which "the refuse of animal and vegetable matters of all kinds, the dirty water from the washing of clothes and of the houses are all poured and they stagnate and putrefy." 71 The unsanitary conditions resulting from overcrowding and poor sanitation created an environment where a number of fatal epidemic diseases could thrive, as illustrated by the great cholera epidemics that raged throughout the cities of England at the height of the industrialization. 72

Public concern over cholera manifested into large public rallies, and riots were common in nineteenth-century England. 73 While it is difficult to estimate the riots' effect on eventual government action, Anthony Wohl has argued that the cholera epidemics of the 1830s, and the public response they elicited, played a large role in sufficiently alarming the central government of England to the need for the government to direct public health in local communities. 74 As a result, some of the initial regulations governing environmental conditions were enacted as Public Health Acts giving guidelines to the local communities in the areas of sanitation, waste disposal, water supply, and drainage. 75

Thus, public concern, evidenced through public demonstrations, spurred government regulation of the environment. During this early period of industrialization one could begin to discern the inextricable nexus between environmental degradation caused by unmitigated industrialization and the need to address public concern associated with that degradation. The connection between human health and environmental management is the primary force behind future environmental standards and regulations.

2. Air Quality

The factory system that arose during the Western period of industrialization encouraged the centralization of production near urban centers. 76 The noxious fumes arising from factory smokestacks created health concerns among many in the growing populations living near factories. Urban dwellers were often forced to shut their windows tightly to keep out the stench and black soot from factories. Such actions merely increased the chances for infection in dwellings where disease was already prevalent. Many citizens in the cities lived under a "black [*407] canopy of smoke" and a constant fog which stung the eyes and inflamed the lungs. 77 In England, local urban health authorities reported hundreds of incidents of nausea, vomiting, bronchial and respiratory complaints, poor digestion, and lack of appetite to the Public Health Boards, which began linking such ailments to air pollution. 78 Several local citizen groups emerged to pressure the government to regulate pollution and promote cleaner air. 79

3. Water Quality

Industrialization and urban growth also polluted the rivers. The need to discard factory and human waste and the population growth of the industrialized city led to the development of sewerage systems that pumped large amounts of untreated sewage into the rivers of newly industrialized nations. One official report described pollution in the Thames in England:

> Throughout the whole course of the river . . . fouling of the water by sewage from cities, towns, villages, and single houses generally prevails. The refuse from paper mills [and] tanneries . . . passes into the stream.

There is no form of scavenging practised for the surface waters of the Thames, but carcasses float down
the stream until wasted by corruption. The river re-
ceives unchecked the whole of the pollution, solid and
fluid, of the district; and this the same water, after it
has been so polluted, is abstracted, sand-filtered and
pumped into the Metropolis for domestic use.80

Industrial waste from factories added large quantities of metallic
salts, lime, and chemical washes that combined to create a poisonous
mix of industrial solvents sufficient to kill fish and infect thousands of
citizens living and working downstream. In Britain, under pressure from
numerous citizen groups, the government formed the Royal
Commission on River Pollution in 1867. 81 The Commission made
numerous recommendations regarding the shape that future legislation
should take regarding water pollution. 82

By the 1870s, increased environmental degradation, growing public
anger over the effects of that degradation, and a largely pro-reform
mindset resulted in a major piece of governmental regulation in the
industrial era-the Public Health Act of 1875. 83 Though the Public
Health Act did not strictly regulate the environment, it did recognize
that the various forms of epidemic and endemic diseases afflicting
industrial cities were directly caused by atmospheric impurities [*408]
produced by both decomposing animal and vegetable substances, and
the filthy conditions caused by overcrowding. 84

The Act was far from comprehensive, and even in its day was
criticized for being, "so little in the interests of the public and so vastly
in the interests of polluters." 85 Even so, the Act was a response to
growing public dissatisfaction with unchecked industrialization and
served as an important precedent for future regulation. For the first
time, government was charged, both nationally and locally, with the
responsibility to take appropriate measures aimed at improving the
quality of the water supply and eliminating waste and refuse. 86

Two forces converged during the Industrial Revolution that had a
profound effect on future generations of environmentalists. First, the
environmental, social, and political consequences of largely unregulated
industrialization became apparent to policy makers and affected citizens
on a scale not possible until the event of mass industry. Second, governments began to realize that neither market forces nor the positive

aspects of industrialization would, alone, mitigate industrialization's negative consequences. Additional regulation was needed if countries were to make successful use of industry. Yet, the scope of regulation and the role the public would play in creating or enforcing any regulatory scheme was still undecided. The Public Health Act in Britain was limited in scope and envisioned only a modest role for the central government, leaving the bulk of regulation a local issue.

C. STAGE THREE: THE U.S. REACTION -- CENTRALIZING THE REGULATORY PROCESS

In the United States, too, the environmental consequences of unregulated industrial development were generating concern among many citizens who by the early twentieth-century began to clamor for some form of government regulation. As in Britain, environmental reform was first a response to obvious health hazards and focused mainly on legislating public health and sanitation at the local level. 87 However, as industrialization continued to expand and the environmental effects of that industrialization became more apparent, a number of factors began to affect the shape that environmental regulation would take in the twentieth-century.

First, the environmental movement became increasingly tied to the progressive movement of the early twentieth-century, a movement which generally sought to reform the ill social effects of industrialization. The progressive movement provided an organized structure in which to house a burgeoning environmental movement, which, unlike its counterpart in Europe, aimed at reforming more [*409] than just sanitary conditions for public health purposes. Martin Melosi argues that by the 1890s, what was once merely a concern for public health had "expanded into a broader environmental perspective, indicating that reformers were beginning to see pollution not merely as an irritant but as an unwanted by-product of industrialization." 88 Fusing with the progressive movement helped to propel the environmental movement into the mainstream of social movements in the United States in the early twentieth century. 89

While the progressive movement popularized and energized the environmental movement, carrying it beyond discussions of nuisance and public health, public calls for regulation were still very local in nature.

This was partly due to the fact that environmental science lagged behind growing environmental concern. Pollution problems were seen as local isolated cases, and reformers did not understand the broad inter-relationships among multiple pollutants. For the average reformer, the goal was to eliminate pollution, not to regulate the source. 90 Many of the Progressive Era environmental movements, though popular, were led by technical professionals working within local municipalities. There was a great belief in the ability of technical experts to develop strategies to fight pollution, and the public as a whole placed a large degree of faith in this process. 91

After World War II, a combination of forces drove the impetus for the modern, centralized, environmental regulatory scheme in the United States. Great strides in the science of ecology revealed the interlocking relationships among population growth, industrialization, resource use, and pollution and demonstrated that environmental degradation was not merely a local problem. 92 The new science, along with a bevy of popular and persuasive books and social activism, fueled and strengthened the conservation-based environmental movements already begun during the Progressive Era. Finally, several well-publicized environmental disasters convinced middle America that local remedies would not be enough to stem a rising environmental crisis. 93 The environmental movement was growing and, by the 1960s, hundreds of environmental groups emerged demanding more centralized federal regulation of industry.

By 1980, the major pieces of U.S. environmental regulation had been enacted, [*410] and the federal era of environmental protection had emerged. 94 These comprehensive statutes established the modern ground rules for environmental protection by mandating environmental impact assessment by public and private entities and establishing controls on air and water pollution, toxic substances, and hazardous wastes.

III. THE IMPORTANCE OF PUBLIC PARTICIPATION RIGHTS TO MODERN ENVIRONMENTAL REGULATORY SCHEMES IN THE INDUSTRIALIZED WORLD

The modern environmental laws of the United States and the European States reflect a policy that encourages all types of active public participation in all aspects of environmental regulation. 95 Under the current environmental regulatory scheme of the United States, public participation mandates are ubiquitous. In fact, the EPA has recently launched a number of programs aimed at expanding the role of the public in the regulatory process in a way that emphasizes the need for a consensus-based holistic approach to environmental decision-making. 96 This section will detail the public participation provisions within the predominant environmental regulatory statutes in the United States as well as within the EPA's recent policy initiatives. For comparative purposes, this section will also highlight high-level environmental public participation initiatives in the European Union. The purpose for highlighting these provisions is to advance two goals: (1) to point out that environmental public participation rights have become crucial to environmental regulation and development in industrialized nations, and (2) to set forth the template for environmental public participation rights by examining the Western incarnation of those rights, a model that is currently being used by transitional States throughout the world.

A. PUBLIC PARTICIPATION MANDATES WITHIN MAJOR ENVIRONMENTAL ACTS OF THE UNITED STATES

From the traditionally reactive authorization of citizen-suit provisions to the more proactive participatory mechanisms aimed at public involvement in the policy formation stage, U.S. environmental regulatory law envisions a broad participation regime.

1. The National Environmental Policy Act

Hailed by many as the statute that launched the "environmental decade" of the [*411] 1970s, the National Environmental Policy Act (NEPA) 97 was signed into law in 1970 by President Richard Nixon. 98 In many ways NEPA represents the culmination of an escalating policy race between both parties in Washington, each determined to convince an increasingly alarmed and active public that it was more committed to environmental preservation and sustainable development than the other. 99 Essentially, NEPA codifies the general environmental policy of the United States by setting forth a number of broad principles and goals designed to ensure environmental responsibility among the agencies of the federal government. 100 Central to the statute is the policy statement that the federal government must use "all practicable means and measures . . . to create and maintain conditions under which man and nature can exist in productive harmony, and fulfill the social, economic and other requirements of present and future generations of Americans," which has become standard phraseology for sustainable development. 101 Under NEPA, each federal agency is responsible for implementing the Act's requirements. However, NEPA also created a central agency, the Council on Environmental Quality (CEQ), to coordinate compliance among the agencies and to conduct general rulemaking for the implementation and enforcement of the Act. 102

To achieve its broad goals, NEPA instructs federal agencies to consider the likely environmental effects of their activities by conducting environmental impact statements (EIS). NEPA's EIS provisions require that all federal agencies:

> include in every recommendation or report on proposals for . . . major federal actions significantly affecting the quality of the human environment, a detailed statement by the responsible official on: (i) the environmental impact of the proposed action, (ii) any adverse environmental impacts which can not be avoided should the proposal be implemented, (iii) alternatives to the proposed action, (iv) the relationship between short-term uses of a man's environment and the maintenance and enhancement of long-term productivity, and (v) any irreversible and irretrievable commitments of resources which would be involved in the proposed action should it be implemented. 103

The courts have interpreted these EIS obligations to have a broad reach, and as a result, agencies devote a lot of time and economic resources to developing EISs.

Congress responded to public concern and demands for greater participation by including various mechanisms in NEPA to encourage the public's participation [*412] in helping develop and implement national environmental policies. 104 First, once an agency prepares an EIS, it must publish the EIS in the Federal Register for public review. 105 The public then, under the procedures set forth in the Administrative Procedure Act (APA) 106 , has an opportunity to comment on the EIS. Second, NEPA authorizes the CEQ to establish and seek assistance from a Citizen's Advisory Committee on Environmental Quality, a group composed of several academics, nonprofit organizations, and citizens. 107 Finally, NEPA provides the EPA with funding to make grants to nonprofit citizen groups to support and encourage participation. 108

The CEQ regulations further stress public involvement in almost every aspect of NEPA implementation, extending from notice and comment procedures to redress in the courts for NEPA non-compliance. 109 In addition, the Act requires individual federal agencies to promulgate their own regulations for implementing NEPA in a way that supplements the CEQ regulations. 110 Taken together, these provisions ensure a strong commitment by all federal agencies to involve the public in government decisions regarding projects impacting the environment.

2. The Clean Water Act

Centuries of unabated industrialization had taken their toll on the fresh water supply of the United States by the middle of the twentieth century. 111 By 1970, most studies estimated that only thirty to forty percent of assessed waters were safe for fishing or swimming. 112 Faced with almost daily stories regarding the adverse health consequences of pollution levels in the water and incessant public calls for greater regulation of industrial polluters, Congress took action in 1970 with the Clean Water Act (CWA), 113 the most comprehensive source of federal regulatory authority to control water pollution. 114 The CWA institutes a permit system whereby legal dischargers are limited, by permit, in the amount of effluent [*413] they may discharge. 115 The CWA also prohibits all unpermitted discharges of pollutants from "point sources" into the "waters" of the United States. 116 Under the Act, states must plan for control of pollution from other nonpoint sources not covered by the permitting process. 117 The CWA has been described as "breathtakingly ambitious" in that the Act's broad mandates impose upon the nation's regulatory framework the unenviable task of wholly eliminating the discharge of illegal amounts of pollutants into the nation's navigable waters. 118

Not surprisingly, public participation plays a prominent role in creating, implementing, and enforcing provisions of the CWA. As with

NEPA, the CWA mandates broad public input, reaching through notice and comment proceedings into the development and enforcement of any regulation, effluent limitation, or agency plan. 119 Further, EPA's regulations implementing the CWA afford the public the opportunity to participate, through advisory groups, in the production of policy guidance documents and citizen guide handbooks. 120 Citizens also routinely participate, through notice and comment proceedings, in the issuance, modification, and enforcement of permits under the National Pollution Discharge Elimination System (NPDES). 121 The EPA and relevant state agencies must not only provide the public the opportunity for input, but they must also promulgate regulations that set up specific guidelines for public participation. 122 This requirement creates a stable structure around the participation process and ensures a minimum level of effectiveness.

In addition to these traditional avenues of public participation, the EPA has recently initiated a number of programs designed to garner even greater public input with regard to the CWA. For instance, the agency has issued a new management strategy that seeks to implement source control measures for nonpoint sources of water pollution. 123 The strategy encourages broad integration of public input and promotes greater use of the Internet and other technology processes as a way to disseminate information and promote public input in environmental decision-making. 124

3. The Clean Air Act

By the middle of the twentieth century, exposure to unhealthy levels of air pollutants such as carbon monoxide, lead, and ozone, along with more hazardous [*414] air pollutants such as mercury and benzene, caused adverse health conditions among many U.S. citizens. 125 There were thousands of documented cases concerning respiratory tract problems, eye and throat irritation, lung disease, retardation, and leukemia among citizens of all races and backgrounds.126

Against this backdrop, the Clean Air Act (CAA) 127 amendments, enacted in 1970, sought to broadly regulate releases of specific types of harmful air pollution. 128 Like the other major environmental statutes created in the early 1970s, the CAA mandated that the EPA, in combination with the states, create and enforce regulations aimed at an ambitious program of pollution reduction.

Similar to the CWA, the CAA allows the public to participate broadly in the creation, implementation, and enforcement of its provisions. The Act provides for public notice and comment to related consent orders and settlement agreements between polluters and the EPA. 129 The Act also mandates public notification of violations of national ambient air quality standards. 130 Further, citizens have an opportunity, under the provisions of the Act, to participate in the EPA's rulemaking procedures regarding implementation and enforcement of the CAA. 131 Finally, the public has the opportunity, through notice and comment proceedings, to participate in the creation and implementation of a state's plan for how to implement the national standards, and citizens have full access to nearly all information submitted by owners and operators of existing, modified, and new sources of air pollution pertaining to whether those owners and operators remain in compliance with the state's plan and with the CAA generally. 132

Some critics have called into question the success of both the CAA and the CWA at implementing broad participation rights both regionally and nationally. 133 However, there is evidence that localized public participation initiatives have garnered attention. For example, in EPA Region 10, the EPA's Water Division published a document called the EPA Lakewalk Manual: A Guidebook for Citizen Participation. 134 This document was designed to empower local participation by teaching concerned citizens how to visit lakes in the region to [*415] compile water quality data. 135 The manual also instructs those citizens on how to turn over the collected data to the EPA or to environmental groups with the capability to analyze the data to determine whether the water quality meets CWA guidelines. 136 The pamphlet includes worksheets and template forms to aid citizens in data recordation.

In another example of successful local participation, the EPA has launched a series of initiatives designed to garner public input in local transportation and economic development planning aimed at reducing detrimental air emissions. 137 These programs are designed to involve the local public earlier in the planning process and to arm citizens with the information necessary to effectively participate in environmental decision-making. 138 The EPA has made it a priority to enhance the public's education, thereby enhancing the public's ability to participate, by providing credible information and technical assistance to local citizens who want to develop alternative strategies for consideration by local planning boards. 139 This meaningful participation has had success in empowering ordinary citizens, in addition to traditional environmental groups, to affect real environmental policy at the grassroots level.

Perhaps most significantly, both the CAA and the CWA provide citizens with the power to enforce the Acts' provisions through use of the citizen suit. 140 Citizen suits permit citizens to act as "private attorneys general" and to bring legal actions to force the EPA to enforce provisions of the environmental statutes. 141 These suits are brought on behalf of the public at large and typically seek to compel action rather than to vindicate purely personal loss. Under the provisions, citizens may bring two types of suits: (1) "enforcement suits" against violators to enforce the requirements of the environmental law in question; and (2) "action-forcing suits" against public officials for an alleged failure to perform "nondiscretionary" duties under the environmental law in question. 142 Citizens bringing suits under these provisions can typically seek injunctive relief, civil penalties, or both. Congress injected the right to bring citizen suits into the CWA and the CAA to guard against the possibility that "special interest creep" in the agency might at times impede the agency from acting to enforce the provisions of the Act. 143 The lawsuits also allow citizen groups to act as the eyes and ears for [*416] the EPA when the agency experiences periods of budget shortfalls. 144 While citizens face a number of constitutional limitations and statutory requirements (both substantive and procedural) in bringing lawsuits under these Acts, the presence and availability of citizen suits as an enforcement mechanism contribute to the already broad arsenal of environmental participatory powers garnered by the public.

4. The Comprehensive Environmental Response, Compensation, and Liability Act

Perhaps more than any other major piece of environmental legislation, the Comprehensive Environmental Response, Compensation and Liability Act (CERCLA) 145 came about as a direct result of growing public concern and dissatisfaction with the unmitigated consequences of unregulated industry practices. 146 CERCLA is arguably the most comprehensive legislative approach to federal environmental protection. 147 CERCLA authorizes the EPA to spend monies from a "Superfund" (funded initially through a tax on chemical feedstocks) on operations for short-term removal of hazardous substances, longer-term remediation, or both, depending on conditions set forth in the Act. 148 The federal government can then bring an action for cost recovery against any or all potentially responsible parties as designated under the Act, whether or not those parties were ultimately responsible for the release of the hazardous substance. 149

Like the other major environmental acts, CERCLA contains provisions creating broad public participation at all levels. Most of the participation provisions are aimed at ensuring the locally affected public plays a crucial role in the cleanup of identified "Superfund" sites located in their communities. 150 For example, the Act mandates a process by which the public can participate in the development of administrative records related to removal and remediation actions. 151 This ensures the local public has a hand in supervising the creation of the only official evidentiary record an administrative court may review upon a legal challenge brought before it. 152 In addition, the public has a right to judicial [*417] review of any regulation promulgated under the Act, and the public must be given notice and chance to comment during public meetings on the scope and effect of all removal actions, including alternative plans. 153 The EPA is required to respond to any significant comments and/or new information submitted during these public meetings. 154 Like the CWA and the CAA, CERCLA ensures that citizen participation in the public meetings is meaningful by offering citizens technical assistance grants designed to aid public understanding of complex clean-up concepts. 155 Once a final clean-up plan and remedy is selected, CERCLA mandates public notification regarding any changes from the original plan. 156 Finally, CERCLA provides for citizen suits against the EPA for failure to perform all nondiscretionary duties under the Act. 157 However, this provision is narrower than those under the CWA and the CAA and has been interpreted by the courts to delay such suits until the remediation is undertaken. 158

B. THE FUTURE OF ENVIRONMENTAL PUBLIC

PARTICIPATION IN THE UNITED STATES: MOVING BEYOND NOTICE AND COMMENT PROCEEDINGS

Recently, the EPA has expanded public participation beyond pure statutory mandates by encouraging a public role in consensus building. The EPA's rationale for its consensus building efforts is that public participation is more effective when key public stakeholders are included in the environmental decision-making process at the earliest possible time and when public stakeholders' opinions are given equal weight in relation to all other stakeholders, such as lobbyists, government officials, experts, and non-governmental organizations (NGOs). 159 Two notable examples are found in recent EPA initiatives calculated to: (1) build holistic integration of environmental decision-making based on consensus building, and (2) include environmental justice protections to existing permitting processes that would ensure the equitable treatment of minority communities by requiring their participation in decisions that affect their environment.

The consensus-building approach, to be effective, requires that all interested parties develop implementation proposals for the initiation of a project affecting the environment or the resolution of an environmental issue. The merits of the [*418] proposals are discussed and debated in a series of roundtable sessions moderated by neutral facilitators, and a final solution is only accepted when consensus, not total agreement, is reached. 160 Every participant must agree that the negotiations were conducted in good faith and that all issues were heard and resolved within the limits imposed by law and economic feasibility.161

An example of this holistic consensual approach to environmental decision-making can be found in EPA's "Green Communities Program," launched in 1997, 162 which encourages local communities to work towards a "shared vision" of sustainable development. 163 The EPA helps communities to work toward a shared vision through a five-step program, based largely around the concept of consensus building. 164 The EPA shares information liberally with communities seeking to implement such programs to enable all the stakeholders to conduct meaningful discussions and create lasting, effective policies based on consensus. 165

Recognizing that poor and minority communities often fail to take advantage of traditional public participation avenues for a variety of reasons, the EPA, pursuant to an executive order, 166 has taken steps to ensure that all citizens have a voice in decisions that affect their environment. The agency has devised an Environmental Justice Checklist that effectively forces the integration of environmental justice considerations into public participation programs. 167 The EPA's injection of environmental justice concepts into its regulations represents the acknowledgement that public participation in environmental decision-making requires inclusion of all public actors to be both fair and effective. Failure to include all actors from the onset of a decision inevitably leads to future conflict if the concerns of affected stakeholders are not addressed.

This approach to environmental decision-making is similar to the Collaborative Compact Model that has been successfully used to create binding industry and community partnerships in the United States. Based upon mutual understanding of needs and priorities, interested parties can develop effective and meaningful rules governing the management of the environment. The Collaborative Compact Model "is based on the belief that there is a clear parallel between [*419] rational regulatory promulgation and reasoned environmental debate." 168

Scholars and practitioners have levied several criticisms against the relevant provisions regarding environmental public participation. 169 They argue that while the current slate of U.S. environmental laws may have put an effective participatory structure in place, the public is still dispossessed of the necessary traits to effectively participate in that structure. 170 Many also argue that the goals of public participation are relatively vague and often conflict with one another. 171 For instance, public participation is said to empower the individual and emphasize direct interaction with and access to decision-makers. These concepts conflict with existing predominant theories of collectivism and republicanism, which are deeply ingrained in modern Western political thought. 172 It is also argued that public participation processes are costly and time consuming and run counter to concepts of efficiency and control, two key rationales for administrative law. 173

With regard to specific acts, critics argue that while the CAA, the CWA, and CERCLA allow stakeholders to participate in all aspects of environmental decision-making under their purview, often they do not do enough to locate key stakeholders, and when they do, they do not go far enough towards helping stakeholders understand the complexities and technical nature of the environmental documents. 174 Finally, many argue that even where the major acts provide for early participation and make available and understandable key documentation, citizen participation fails its objectives because citizen groups (and especially individual citizens) do not have the resources to compete with other stakeholders whose interests may run counter to their own. 175

These are valid considerations, and they are the subject of several important scholarly articles. However, notwithstanding the ultimate effectiveness of the participation provisions and mandates in any part or all of the major acts in U.S. environmental law, the fact remains that the public has come to play a key role in environmental decision-making and cannot be extricated from the process.

[*420]

C. ENVIRONMENTAL PUBLIC PARTICIPATION IN THE EUROPEAN UNION: AARHUS AND BEYOND

This subsection highlights the major trends in environmental public participation throughout Europe by examining the pertinent provisions of the Aarhus Convention, a far-reaching multilateral agreement focused on public participation rights. 176 The purpose of this subsection is to expose a thread of continuity among European Union, European, and U.S. law regarding the great extent to which environmental public participation rights, and the elaborate legislative mechanisms to secure those rights, have surfaced and now play a crucial role in environmental decision-making throughout most developed regions of the world.

The first fifteen member States of the European Union, and the European Community itself, have created environmental regulatory schemes containing significant provisions allowing for broad public participation in environmental decision-making. Like the United States, the shared historical experiences of Europeans have led to the recognition that the public must play a crucial role in implementation and enforcement of policies that affect the air, land, and water around it. Building on this tradition of environmental public participatory rights, the Aarhus Convention sought to create a multilateral agreement, negotiated and signed under the auspices of the United Nations, to ensure that all countries guarantee some form of participation in environmental decision-making. 177 Forty European States, including the first fifteen member States of the European Union and the European Community, have signed the agreement. 178 Several EU member States have already ratified the agreement, and most others are currently working internally to secure the support necessary for ratification. 179

The Aarhus Convention aims to ensure basic public participatory rights in environmental decision-making. The agreement binds States to establish a minimum baseline structure upon which they can build three core systems of participatory rights: (1) the public's right of access to information, (2) the public's right to a forum to participate, and (3) the public's right of access to the justice system. 180 These principles establish the basic foundation of environmental public participation rights and have been developed for some time in academic literature. However, they were first embodied as legal principles in the Rio Declaration in 1992. 181 Most developed European States already provide these [*421] rights to their citizens. However, the Aarhus Convention ensures that these rights never fall below a minimum threshold. 182 In this sense, Aarhus represents a reaffirmation by Europeans that the public should play an important role in decisions that affect the environment. Moreover, Aarhus seeks to bring the participatory practices in the transitional States within Europe into balance with the more developed European States and the European Union generally. 183

Aarhus assigns the same importance to environmental rights as to human rights and calls on its signatory nations to work toward the incorporation of environmental rights to that degree (i.e., by giving environmental rights constitutional guarantees). 184 While most of the parties to the agreement have stopped short of elevating environmental public participation rights to the level of constitutionally guaranteed human rights, all of the signatories have imposed upon themselves the obligation to provide all citizens with access to information and the opportunity to participate.

Most significantly, Aarhus represents an earnest attempt to comply with one of the core principles of the Rio Declaration-that environmental public participation rights are not just legal rights in a democratic society, but also are crucial human rights essential to successful sustainable development and to securing the rights of future generations. 185 The concept is critical to understanding the evolution of environmental public participation rights from rights that are merely given by the government to rights that cannot be taken away. Such advanced notions of participation as a human right are gaining ground globally and provide a glimpse into the future scope of public participation rights.

IV. TOWARD AN INTERNATIONAL HEGEMONY: ENVIRONMENTAL PUBLIC PARTICIPATION RIGHTS, SUSTAINABLE DEVELOPMENT, AND THE DEVELOPING WORLD

Two forces play crucial roles in pushing transitional States to implement environmental public participation rights. First, a number of important multilateral agreements, both international and regional, have given import and legal substance to the concept of public participation in environmental decision-making. These agreements have helped create sound legal principles to which many transitional States look for guidance in creating and implementing their own mechanisms. 186 Second, there is growing internal pressure on transitional States, rooted in public concern for the environment that acts as a driving force for the implementation of environmental public participation rights.

[*422] This section demonstrates how the twin concepts of sustainability and participation have deeply influenced binding regional agreements and international aid policies for development. Consequently, transitional States, bound by formal mandates, are implementing environmental public participation rights to an unprecedented degree. This section also describes how overwhelming internal pressure from concerned citizens forces transitional States to involve the public in environmental decision-making.

A. INTERNATIONAL ENVIRONMENTAL AGREEMENTS

Environmental public participation is a relatively recent concept within the recognized framework of broad international initiatives. 187 In fact, from 1940 to 1970 there were no explicit environmentally-oriented international initiatives. 188 Instead, early international initiatives focused on concepts such as human rights and equity. Yet, these instruments played an important role in laying the foundation for future international discourse on environmental public participatory rights. For example, the Universal Declaration of Human Rights provides that "everyone has the right to take part in the government of his country, directly or through chosen representatives." 189 Article 19 of the Declaration provides also that "everyone has the right to freedom of opinion and expression; this right includes freedom to . . . seek, receive, and impart information and ideas through any media." 190 Likewise, the International Covenant on Civil and Political Rights provides that every citizen "shall have the right and the opportunity . . . (a) to take part in the conduct of public affairs, directly or through freely chosen representatives." 191 Thus, while these instruments do not focus specifically on public participation in environmental issues, they do set forth principles that form the basic prerequisites to public participation and serve to acknowledge public participation as a basic human right, deserving protection and encouragement. 192

In the 1970s and 1980s, two events resulted in a growing number of international environmental initiatives containing specific provisions that focused on the right to public participation in environmental decision-making. First, a series of environmental crises (e.g., the toxic contamination resulting from the 1978 Love [*423] Canal disaster and the 1984 Union Carbide explosion in Bhopal, India) forced leaders and organizations throughout the world to take notice of the global scale of environmental degradation. 193 This heightened awareness stimulated a number of international meetings on the environment and provided the catalyst for the global movement towards the concept of sustainable development. 194 Second, increasingly popular and powerful environmental NGOs were permitted a place at the table during these meetings, and they forced the international community to include public participation rights as part of sustainable development. 195

1. Sustainable Development

The concept of sustainable development emerged in 1972 with the commencement of the U.N. Stockholm Conference on the Human Environment. 196 This conference is generally recognized as marking the first large-scale international meeting linking the effects of human activities with environmental degradation. 197 Stockholm's stated goal, among other things, was to highlight the extent of the degradation resulting from pollution and resource destruction, and the conference was the first to acknowledge that social well-being was intimately related to environmental well-being. 198 The Declaration acknowledged that economic development was necessary for improvements in the living standards of the world's poorer and undeveloped States and proposed a number of principles to ensure such development would be sustainable. 199 Perhaps the most successful accomplishment of Stockholm is that it placed environmental concerns at the forefront of the international legal agenda for the next three decades and spawned a number of important international agreements regarding sustainable development. 200

In 1987, the U.N.-sponsored World Commission on Environment and Development (Brundtland Commission) released Our Common Future, perhaps the most influential expression on sustainable development to date. 201 The Brundtland Commission, like Stockholm, posited that economic development was necessary for human development but stressed that current patterns of development and consumption (in both the developed and developing areas of the world) must [*424] change course to fit within the planet's ecological limits. 202 Our Common Future stressed that environmental concerns and economic concerns were related and that both must be attacked through a series of initiatives aimed at poverty reduction, education, conservation, and, most importantly for this paper, public participation. 203

Global attention to sustainable development peaked at the 1992 U.N. Conference on Environment and Development in Rio de Janeiro (Rio). 204 The size and scope of Rio speaks to its significance. The conference brought together representatives from 179 States and included the Earth Summit, the largest-ever meeting of world leaders. 205 Taking the helm from the Brundtland Commission, the conference aimed to address, authoritatively, what many considered to be an emerging triumvirate of crucial problems: environmental protection, social progress, and economic development. 206 In the end, Rio produced two international agreements, two statements of principles, and a major action agenda on worldwide sustainable development, thus laying the groundwork for States around the world to incorporate the concept of sustainable development into their binding regional agreements and domestic legislation. 207

2. Injecting Public Participation into Sustainable Development: The Role of the Non-Governmental Organization

One of the forces driving the notion that public participation should play an important role in the evolving concept of sustainable development was the growing prominence of NGOs. 208 The public was not involved in the processes that produced early environmental initiatives such as Stockholm; there was simply no protection of the public interest other than through the State representatives. The recognition of public involvement had to surface in the processes that produced the initiatives before it could appear in the text of the initiatives themselves. Once the United Nations allowed NGOs to take part in environmental conferences, the NGOs pushed for provisions that would allow for greater public participation and promote greater public awareness and education. 209

Consequently, the international environmental instruments that emerged throughout the 1980s and 1990s reflected this new spirit of participation. For example, the World Charter for Nature, a resolution of the U.N. General [*425] Assembly, explicitly imposes a duty upon the public to act in accordance with their national legislation and provides that "all persons . . . shall have the opportunity to participate . . . in the formulation of decisions of direct concern to their environment." 210 Several environmental agreements have since followed suit with similar provisions.211

Thus, the twin concepts of sustainability and participation combined to make environmental public participation rights a key feature in the modern international environmental landscape. By the time of the Rio Conference in 1992, global support for environmental public participation reached its zenith of expression. 212 Addressing public participation specifically, Principle 10 of the Rio Declaration states:

> Environmental issues are best handled with the par-
> ticipation of all concerned citizens, at the relevant
> level. At the national level, each shall have appropri-
> ate access to information concerning the environment
> that is held by public authorities, including informa-

tion on hazardous materials and activities in their communities, and the opportunity to participate in decision-making processes. States shall facilitate and encourage public awareness and participation by making information widely available. Effective access to judicial administrative proceedings, including redress and remedy, shall be provided.213

The Rio Declaration essentially codified the basic and familiar concepts underlying effective public participation: (1) access to information, (2) right to participate in decision-making, (3) right to judicial enforcement, and (4) remedy. 214 With the Rio Declaration, public participation in environmental decisionmaking became a legitimate part of the discourse of international environmental law and a legitimate policy expression to which many transitional States would strive to achieve. 215 Furthermore, the philosophy of the Rio Declaration would come to permeate all subsequent major international environmental initiatives that now almost universally affirm the concept of public participation. 216

The importance of broad international initiatives such as the Rio Declaration lies not in any particular legal obligation set forth by the initiative itself; rather, such initiatives are important because they provide a framework for an operational [*426] right that reflects international consensus. 217 Furthermore, such broad instruments often form the basis for regional agreements, many of which not only set forth more specific obligations but are enforceable as multilateral and bilateral treaties. Because Rio has tied public participation to the concept of sustainable development, countries are implementing participatory processes as a part of their overall desire to implement sustainable development initiatives. 218

A number of regional initiatives also incorporate public participatory rights into their environmental agreements. For example, Asian and European (ASEM) partners have developed both binding and non-binding regional agreements focusing on environmental conservation and sustainability. 219 These agreements contain provisions that promote the enhancement of public environmental awareness by developing and implementing education programs and facilitating the sharing of environmental information among government, industry, and citizen groups. 220 Further, ASEM has embarked, at the ministerial level, on a non-binding Asian initiative that would seek to implement public involvement in environmental issues on the level of that sought by Europeans under the Aarhus Convention. 221

Likewise, the Arab Ministerial Conference on Environment and Development has issued an Arab Declaration on Environment and Development and Future Perspectives (Arab Declaration). 222 The Arab Declaration, among other things, stresses that environmental protection and sustainable development require that States give "support" to individuals, local organizations and non-governmental organizations. 223 The Arab Declaration views these groups as representative of popular participation, a concept which must be given "due attention." 224 Under this view, individuals and organizations shall have the "right to acquire information about environmental issues . . . access to data . . . and to participate in the formulation and implementation of decisions that may affect their environment." 225

In Central and South America, the Organization of American States (OAS) has initiated the Inter-American Program of Action for Environmental Protection. 226 This resolution of the General Assembly of OAS places a high priority on public awareness and participation. It directs its members to promote greater education [*427] and information access to the general public and NGOs so that such groups can participate in efforts to improve and protect the environment in the region. 227

This section has provided a glimpse of international environmental agreements, instruments, and initiatives that deal with public participation rights to demonstrate the global evolution of such rights. Public participation rights have evolved in the international environmental legal arena as steadfastly as they have evolved within the environmental legal regimes of the most developed States in the world. Transitional States are increasingly bound to insert the rights created in these agreements into their own domestic legislation.

B. INTERNAL PRESSURE FORCES PUBLIC PARTICIPATION: THE CAMBODIAN EXAMPLE

While international initiatives identify a framework for environmental public participation rights, the driving force for implementation of those rights, especially in a way that binds the State, can only come from strong internal public demand. In responding to internal pressure, transitional States can look to the international framework for guidance on how to properly channel public participation so that public discord does not cause potentially harmful instability.228

The Cambodian example illustrates that where transitional countries neglect to implement sufficient environmental regulatory control and to include the public in the regulatory processes affecting the environment, the public response to environmental degradation can be overwhelmingly destabilizing. Faced with such internal pressure and agitation for greater regulatory control over industrial growth, as well as calls for improved public participation, transitional States like Cambodia are increasingly turning to the international framework for guidance.

Cambodia, like many transitional States, suffers from a past marred by political violence, a lack of financial resources, and a centralized government whose actions and officials are not often subject to oversight or public scrutiny. There is a wealth of natural resources in the country, and the government is currently actively pursing a course of infrastructure development. However, Cambodia currently lacks an effective environmental regulatory structure, and the affected public in Cambodia has very few outlets for participation in environmental decision-making. 229

[*428] In 1998, 4000 tons of hazardous toxic waste containing high levels of mercury were dumped illegally in Cambodia. 230 The waste originated at a Taiwanese petrochemical company. 231 Two people died from direct exposure to the waste, and the ensuing backlash sparked mass rioting and hysteria among local citizens. Fearing contamination, upwards of ten thousand residents fled the area. 232

As the weeks went on, allegations were investigated as to whether the Taiwanese chemical company gave several members of the Cambodian government U.S. $ 3 million in bribes to ignore its illegal dumping activities. 233 The illegal disposal of the hazardous waste and the associated political scandal illuminated the near-total inadequacy of Cambodia's environmental laws. There were no records of the dumping, few reporting procedures, and little regulatory process by which anyone could be held accountable. 234 234 Feeling powerless and frustrated with their government's reluctance to act, many Cambodian citizens throughout the country took to the streets and launched massive protests. 235 Subsequent government crackdowns on these riots only fueled more violent resistance, resulting in further turmoil and even death. 236 Within a matter of weeks, ever-larger public crowds were demanding apologies from the key actors, even Taiwan, and calls to overthrow the current regime were becoming more vocal. 237 In a short time, Cambodia was facing a real crisis, largely due to public anger over inadequate environmental regulatory processes and a lack of formal public participation.

Ultimately, the government arrested around one hundred public officials, demanded an apology from the Taiwanese chemical company, and required the company to clean up its waste. 238 Fearing future backlash, the Cambodian government is now attempting to change many of its existing environmental laws to reflect greater controls on waste management and create more thorough reporting mechanisms for hazardous waste dumping. 239 The government is also exploring ways to provide the public with greater dissemination of information pertinent to environmental decision-making, including the initiation of several [*429] programs to educate the public on how to monitor waste. 240

V. ECONOMIC INCENTIVES FOR ENVIRONMENTAL PUBLIC PARTICIPATION

There is another compelling factor driving transitional and developed States to embrace the concept of environmental public participation: it simply makes good economic sense. Capital investors generally recognize that the public's ability to participate in environmental decision-making produces positive economic outcomes essential for successful development. Successful economic markets require a free flow of information. Because environmental public participatory schemes stress the importance of education and access to information, they facilitate the free exchange of information, thereby fueling investment, a crucial component of development. 241

Perhaps most importantly, public participation mitigates economic loss by properly allocating the risks associated with development, including environmental risks, among all the parties involved. 242 As such, transitional States and capital investors are beginning to implement environmental regulatory regimes and public participation as a means of achieving sustainable growth, allocating risks, and attracting investment. 243 This section explores the concept of environmental risk, as faced by transitional States, and how publication participation rights can allocate that risk among interested stakeholders.

A. IMPLEMENTING PUBLIC PARTICIPATION AS A MEANS TO ALLOCATE ECONOMIC RISK IN TRANSITIONAL STATES

To effectively compete in the global economic atmosphere, transitional States must put into place a solid infrastructure. Infrastructure is a foundational element of economic development and includes power generating and distribution facilities, telecommunications networks, irrigation and freshwater systems, waste management and disposal operations, and transportation mechanisms. 244 In the competitive world of transitional States, a well-developed infrastructure provides the advantage critical to attract foreign investment and to retain domestic capital.

There are two certainties transitional States face when attempting to construct [*430] infrastructure on a grand scale: (1) infrastructure projects are extremely costly and usually require foreign direct investment from outside public or private sources, and (2) infrastructure projects will have an impact on the environment of the local communities hosting them and on the transitional State as a whole. In fact, there is evidence that a rapid rate of infrastructure growth (and consequently economic growth) is proportionately related to an increase in environmental degradation. 245

It is often argued that environmental degradation occurs, among other reasons, because transitional States fail to enforce environmental regulations, even where they exist, to lower project costs and attract investors. 246 In effect, States neglect to internalize the costs of environmental degradation in implementing their infrastructure projects for fear that higher short-term project development costs will stifle economic growth, 247 thereby passing on environmental costs to the public as a whole. 248 Transitional States have more economic incentive to use the environment as a form of subsidy for development rather than to impose a regulatory scheme on industry to protect the environment. 249 Under this economic rationale, the local citizenry that ultimately bear the external environmental costs of development is almost completely excluded from the decisionmaking process.

However, as the Cambodian case demonstrates, the long-term costs imposed by a State's failure to enforce its regulations and to embrace public participation in environmental decision-making can be much more significant than short-term economic gain. 250 Environmental degradation is an economic cost, as is unmitigated public angst. Conversely, both the procurement of public "buy-in" and the mitigation of the environmental effects of development confer an economic benefit because they can properly allocate risk among all stakeholders. The actions of the Mekong River Commission (MRC) provide a useful example.

B. CASE STUDY: MEKONG RIVER COMMISSION

The following case study demonstrates how transitional States are moving to implement international standards of environmental public participation rights as a means of alleviating public discord and properly allocating the environmental risk associated with development. 251 The participatory instruments relevant to [*431] the agreement among the States forming the MRC illustrate that there is economic, social, and political utility for States to create and implement effective environmental public participatory schemes.

The Mekong River is one of the world's great rivers and a major income provider for millions of people. 252 For thousands of years, the river's annual flood-drought cycles have been essential for the sustainable production of rice and vegetables on the floodplains and along the riverbanks during the dry season. 253 The Mekong River basin also has great potential for developing hydro-electricity facilities, transportation infrastructure, agricultural innovation, abundant fisheries and a tourism industry, all of which are significant to the social and economic development of the countries in the basin. 254

Given its potential for economic development, there are serious challenges affecting the viability of the river. The current population in the Mekong River basin is over 60 million and is likely to exceed 100 million by 2025. 255 A dramatic increase in population will no doubt serve to increase pressure on the river's natural resources. Also, investors, eager to generate capital, seek to develop necessary infrastructure as quickly as possible and with few restrictions. 256 The need to balance the desire for economic development in the area with the desire to protect the environmental sustainability of the river basin has never been greater. It is crucial that all of the countries within the Mekong River basin cooperate to create economic development that is sustainable, reduces poverty, builds capacity, and protects the environment and ecosystem. 257

Largely due to the influence of Rio and local NGOs, the MRC was reorganized in 1995 as an international river basin authority with the status of an international regional treaty. 258 The MRC is made up mainly of member States from southeast Asia, and has monetary support from donor countries and from the World Bank, the Asian Development Bank and U.N. Development Program. 259 The stated purpose of the MRC is to establish cooperation in all fields of sustainable development, including utilization, management and conservation of the water and related resources of the basin, navigation, flood control, fisheries, agriculture, hydropower, and environmental protection. 260

[*432] More recently, the MRC established the Mekong River Basin Development Plan (BDP) to identify and seek investment for high priority projects directed to achieving sustainable development in the region. 261 The MRC and the donor agencies recognized that the most ambitious infrastructure plans were the ones that affected the most stakeholders and had the greatest potential impact on the environment. However, these ambitious infrastructure plans also held out the most promise for beneficial economic development. Thus, the BDP makes a serious effort to identify the most useful (yet most potentially harmful) projects and to subject those projects to a rigorous plan designed to achieve sustainable development according to the principles set forth in Rio and its progeny. 262

What is most striking about the BDP, for the purposes of this paper, is the agreement's level of commitment to environmental public participation as a crucial element of sustainable development. Arguably, three of the four member countries are not genuinely democratic. They have long histories of under-inclusion of public stakeholders and have traditionally offered few avenues for public participation in government decision-making generally. Yet, when it came to the development of the BDP, the MRC committed itself to an open process whereby "all concerned shareholders will have a chance to participate." 263 The MRC has made good on its promise to include the public. For example, in 1999 the MRC launched a BDP task force to study the ways in which it could advance meaningful public participation in its decisions and has since developed a policy geared towards assuring that will happen. 264 The approach has already been met with some success, and the MRC has become widely recognized for these successful efforts. 265

VI. CONCLUSION

Since the earliest days of organized society, the public has demanded involvement in its environment because of, and typically in response to, adverse epidemiological impacts. The experiences of the MRC and Cambodia demonstrate that environmental public participation rights are necessary for development. Deliberate, effective, and well-reasoned growth can occur only after critical involvement of host community stakeholders and the impacted public. Transitional States, as well as the greater international community, have embraced [*433] the concept of environmental public participation rights, and a growing number of regional multilateral agreements are producing successful outcomes as a result of incorporating environmental public participatory schemes into workable sustainable development plans. Ultimately, such rights cannot and should not be ignored by government or industry.

FOOTNOTES:

n1 Public participation here means, generally, purposeful activities in which citizens take part in relation to government action or decision-making. See generally Nancy Perkins Spyke, Public Participation in Environmental Decisionmaking at the New Millennium: Structuring New Spheres of Public Influence, 26 B.C. ENVTL. AFF. L. REV. 263, 266 (1999). For a definition of public participation in environmental decision-making, see *infra* note 2.

n2 See Neil Popovic, The Right to Participate in Decisions that Affect the Environment, 10 PACE ENVTL. L. REV. 683, 683 (1993). This paper uses the term "environmental public participation rights" in order to capture the entire bundle of rights associated with public participation generally and apply such rights to environmental decision-making. Environmental public participation rights can generally be defined as the rights of the public citizenry within a nation-state to play a role in the creation, implementation, and enforcement of decisions, policy formation, and regulation that affects the environment. *Id.* Most scholars agree that for environmental public participation to be effective it must require, at a minimum, six elements: (1) education about the environment and factors which affect it, (2) access to information (and availability of that information), (3) participation in decision-making process, (4) transparency in the decision-making process, (5) opportunity for post-project analysis and monitoring, and (6) access to justice (enforcement and redress). *Id.*

n3 International law here is taken broadly to mean all legal instruments, principles, understandings, and frameworks that arise from multilateral or bilateral conventions, declarations, agreements, and the general body of customary international law.

n4 See *infra* Part IV.

n5 See *infra* Part IV.

n6 See, e.g., WORLD BANK, The Impact of Environmental Assessment xviii, n.4 (Technical Paper No. 363, 1997); ASIAN DEV. BANK, Environmental Risk Assessment: Dealing with Uncertainty in Environmental Impact Assessment 9-12 (Env't Paper No. 7, 1991).

n7 See generally Thomas Mullikin & Nancy Smith, Community Participation in Environmental Protection, 21 UCLA J. ENVTL. L. & POL'Y 75 (2003).

n8 See *infra* Part III.

n9 Convention on Access to Information, Public Participation in Decision-Making and Access to Justice in Environmental Matters, June 25, 1998, U.N. Doc. ECE/CEP/43, 38 I.L.M. 517 [hereinafter Aarhus Convention]; see *infra* notes 210-14 and accompanying text. The EU came into existence on November 1, 1993 as a result of the Maastricht Treaty. It incorporated but did not replace the EC. The EC is still referred to when discussing functions conferred by the European Community Treaty or its legal effect.

n10 See *infra* Part IV.B.

n11 Popovic, *supra* note 2, at 684 (quoting Henry J. Steiner, Political Participation as a Human Right, 1 HUM. RTS. Y.B. 77, 77 (1988)).

n12 See *infra* Parts II and III.

n13 See *infra* Part III.

n14 See *infra* Part IV.

n15 See *infra* notes 227-37 and accompanying text (Cambodia).

n16 See *infra* Part V.

n17 This is clearly changing. See *infra* notes 238-63 and accompanying text.

n18 In this article the terms "environmental stressors" and "environmental stress" are taken to mean those general factors (e.g., air pollution, water pollution, noise pollution, industrial waste, overcrowding, etc.) associated with the processes of large-scale industrial development. See generally MARTIN V. MELOSI, EFFLUENT AMERICA: CITIES, ENERGY, INDUSTRY AND THE ENVIRONMENT (2001) (discussing the connection between environmental stress and industrial development); ANTHONY WOHL, ENDANGERED LIVES: PUBLIC HEALTH IN VICTORIAN BRITAIN (1983).

n19 This paper assumes that as environmental regulatory schemes develop, the chances for meaningful public participation increase. Therefore, the study of the developmental stages of environmental regulation in industrial society provides also a greater understanding of how environmental public participation has evolved over time.

n20 A fourth element could be added: resources to effect regulation must also be available if regulation is to be accepted. However, resources are not necessary merely for the creation of an environmental regulatory scheme.

n21 ROBERT V. PERCIVAL ET AL., ENVIRONMENTAL REGULATION: LAW SCIENCE AND POLICY 60 (4th ed. 2003).

n22 FRANK J. FROST, GREEK SOCIETY (5th ed. 1997); HENRY C. BOREN, ROMAN SOCIETY (2d ed. 1992); Paul Krugman, Editorial, Salt of the Earth, N.Y. TIMES, Aug. 8, 2003, at A17.

n23 Garbage through the Ages, Wayne County Soil and Conservation District Webpage, at http://www.wastenot.org/articles/garbage_through_ages.html (last visited Feb. 5, 2005).

n24 INSTITUTE OF WASTES MANAGEMENT, HISTORY OF WASTE AND RECYCLING INFORMATION SHEET, at http://www.wasteonline.org.uk/resources/ InformationSheets/HistoryofWaste.htm (last visited Feb. 5, 2005).

n25 *Id.*

n26 Christian anthropocentrism stems from the Christian understanding of creation in which God created the earth and all of the plants, birds, fishes, and other animals on the planet for the benefit of man alone who was made in God's image. No item in the physical creation had any purpose other than to serve man's purposes. This belief, strongly held in the Middle Ages and up to the Scientific Revolution of the eighteenth century naturally stifled a great deal of early scientific curiosity and interest in ecological matters. See generally Lynn White, Jr., The Historical Roots of Our Ecological Crisis, 155 SCI. 1203 (1967).

n27 See generally FRANCES GIES & JOSEPH GIES, LIFE IN A MEDIEVAL CITY (1969); JACKSON J. SPIELVOGEL, WESTERN CIVILIZATION -- VOLUME I: TO 1715, 267-304 (4th ed. 2000).

n28 See GIES & GIES, *supra* note 27, at 161-63.

n29 SPIELVOGEL, *supra* note 27, at 340-42.

n30 *Id.* at 266.

n31 See GIES & GIES, *supra* note 27, at 212.

n32 *Id.*

n33 SPIELVOGEL, *supra* note 27, at 268.

n34 Spielvogel writes that the goods of northern Europe could be traded at fairs in western Europe with goods from southern Europe and east Asia. "Northern merchants bought the furs, woolen, cloth, tin, hemp, and honey. . .and exchanged them for the cloth and swords from northern Italy and the silks, sugar, and spices from the East." *Id.*

n35 *Id.*

n36 See GIES & GIES, *supra* note 27, at 167.

n37 SPIELVOGEL, *supra* note 27, at 272.

n38 *Id.*

n39 *Id.*

n40 GIES & GIES, *supra* note 27, at 77. See also Atmosphere Climate and Environment Information Programme, History of Air Pollution, in ENCYCLOPEDIA OF THE ATMOSPHERIC ENVIRONMENT, at http:// www.ace.mmu.ac.uk/eae/english.html (last visited Feb. 5, 2005).

n41 See, e.g., SPIELVOGEL, *supra* note 27, at 272 (citing an order by the King of England to the town of Boutham requiring rectification of the town's physical condition).

n42 *Id.* (citing city government attempts to require tradesmen and citizens to cart garbage and waste a specified distance from the city walls).

n43 GIES & GIES, *supra* note 27, at 315.

n44 *Id.*

n45 The Black Death of course is the common historical name given to the smaller series of plagues that swept through Europe in the decades after 1340, in some localities causing a mortality rate as high as sixty-five percent. See generally PHILIP ZIEGLER, THE BLACK DEATH (1991); FRANCES GIES & JOSEPH GIES, MARRIAGE AND FAMILY IN THE MIDDLE AGES (1987).

n46 See, e.g., Pistoia: Ordinances for Sanitation in a Time of Morality, in INTERNET MEDIEVAL SOURCEBOOK (Paul Halsall ed., 1996), at http://www.fordham.edu/ halsall/med/pistoia.html (last visited Feb. 5, 2005); Plague and Public Health in Renaissance Europe, Internet History Sourcebooks Project, at http://www.iath.virginia.edu/ osheim/ (last visited Feb. 5, 2005).

n47 See Pistoia: Ordinances for Sanitation in a Time of Morality, *supra* note 46.

n48 SPIELVOGEL, *supra* note 27, at 274.

n49 *Id.* at 276.

n50 *Id.* at 460-84.

n51 See *supra* note 20 and accompanying text.

n52 See MARVIN PERRY ET AL., WESTERN CIVILIZATION: IDEAS, POLITICS & SOCIETY 511-31 (6th ed. 2000); JACKSON J. SPIELVOGEL, WESTERN CIVILIZATION -- VOLUME II: SINCE 1550, 582-607 (4th ed. 2000); MERRY WIESNER ET AL., DISCOVERING THE WESTERN PAST -- VOLUME II: SINCE 1500, 143-78 (4th ed. 2000).

n53 Britain was the first to industrialize in the late eighteenth-century. Enabled in part by town growth and food surpluses, Britain's industrialization brought mechanized factory production, steam engine, coal, and transportation systems to the fold. *Id.*

n54 *Id.*

n55 *Id.*

n56 See ERIC HOBSBAWM, THE AGE OF CAPITAL 29-48 (1975).

n57 *Id.*

n58 See WILLIAM J. DUIKER, TWENTIETH CENTURY WORLD HISTORY 1-10 (1999).

n59 HOBSBAWM, *supra* note 56, at 235.

n60 DUIKER, *supra* note 58, at 7.

n61 *Id.*

n62 HOBSBAWM, *supra* note 56, at 230-50.

n63 See PERRY, *supra* note 52, at 512.

n64 E.g., The population of London doubled between 1800 and 1850. *Id.*; see also WOHL, *supra* note 18, at 3.

n65 WOHL, *supra* note 18, at 4.

n66 *Id.*

n67 SPIELVOGEL, *supra* note 52, at 597 (citing an 1843 report on working class housing by the City of Birmingham).

n68 *Id.*

n69 E. ROYSTON PIKE, HUMAN DOCUMENTS OF THE INDUSTRIAL REVOLUTION IN BRITAIN 313 (1966).

n70 *Id.* at 314.

n71 *Id.* at 343.

n72 WOHL, *supra* note 18, at 118-20.

n73 *Id.* at 119. In 1832, over thirty riots broke out in London, Liverpool, Manchester, and other industrialized cities throughout England. *Id.*

n74 See WOHL, *supra* note 18.

n75 See, e.g., The Public Health Acts of 1848, 1866, 1872, 1875; Infectious Diseases (Notification) Act, 1889, 52 Vict., c.75 (Eng.); Rivers Pollution Prevention Act, 1876, 39 & 40 Vict., c.75 (Eng.) (Available in CHARLES JOSEPH HAWORTH, THE STATE LAW RELATING TO RIVERS POLLUTION (1897)).

n76 See MELOSI, *supra* note 18, at 27.

n77 See WOHL, *supra* note 18, at 208.

n78 *Id.*

n79 *Id.*

n80 *Id.* at 233 (citing an 1867 report by the Royal Commission on the Pollution of Rivers).

n81 *Id.* at 235.

n82 *Id.* at 236.

n83 Public Health Act, 1875, 38 Vict., c.55 (Eng.).

n84 SPIELVOGEL, *supra* note 52, at 599.

n85 WOHL, *supra* note 18, at 248.

n86 *Id.* at 246-247.

n87 See MELOSI, *supra* note 18, at 40.

n88 *Id.*

n89 *Id.* at 213. Major Progressive Era environmental reform movements such as the conservation movement and the urban environmental movement made great strides to educate the urban population on the issues of resource conservation and addressing industrial waste and pollution.

n90 *Id.* at 59.

n91 *Id.* at 214.

n92 See, e.g., PAUL R. EHRLICH, THE POPULATION BOOM (Ballantine Books 1986) (1968); BARRY COMMONER, THE CLOSING CIRCLE: NATURE, MAN AND TECHNOLOGY (Bantam 1980) (1971); Garrett Hardin, The Tragedy of the Commons, 162 SCI. 1243 (1968).

n93 For example, 300 people died in New York City in 1963 due to the existence of concentrated air pollutants, the oil-polluted Cuyahoga River caught fire and burned for eight days in 1969, and 1969 brought the fish-kill in Lake Erie.

n94 See, e.g., National Environmental Policy Act, 42 U.S.C. §§ 4321-4370(f) (2004); Clean Water Act, 33 U.S.C. §§ 1251-1387 (2004); see also PERCIVAL, *supra* note 21, at 88.

n95 See *supra* note 2 and accompanying text.

n96 See *infra* notes 158-73 and accompanying text.

n97 42 U.S.C. § 4231-4347 (2004).

n98 PERCIVAL, *supra* note 21, at 783 (citing Bradley C. Karkkainen, Toward a Smarter NEPA: Monitoring and Managing Government Environmental Performance, 102 COLUM. L. REV. 903, 904 (2002)).

n99 *Id.* at 784.

n100 *Id.*

n101 42 U.S.C. § 4331(a) (2004); see *infra* notes 195-205 and accompanying text.

n102 42 U.S.C. §§ 4342, 4344 (2004).

n103 42 U.S.C. § 4332(c) (2004).

n104 42 U.S.C. § 4331(c) (2004) ("Each person has a responsibility to contribute to the preservation and enhancement of the environment.").

n105 42 U.S.C. § 4332(c)(i)-(v) (2004).

n106 5 U.S.C. §§ 551-596 (2004).

n107 42 U.S.C. § 4345 (2004).

n108 42 U.S.C. § 4368 (2004).

n109 See, e.g., 40 C.F.R. § 1501.7 (2004) (calling for public input in agency "scoping" decisions); 40 C.F.R. § 1502.19 (2004) (allowing for ready public distribution of environmental impact statements to all who request draft copies of the relevant documents); 40 C.F.R. § 1503.1(a)(4) (2004) (requiring agencies to allow for open comment procedures); 40 C.F.R. § 1503.4 (2004) (requiring agencies to respond to comments from concerned parties).

n110 40 C.F.R. § 1507.3(a) (2004).

n111 See *supra* Part II.

n112 PERCIVAL, *supra* note 21, at 570.

n113 33 U.S.C. §§ 1251-387.

n114 PERCIVAL, *supra* note 21, at 581.

n115 *Id.* at 582 (citing § 402 of the CWA).

n116 *Id.* (citing § 301 of the CWA).

n117 *Id.* (citing § 319 of the CWA).

n118 *Id.*

n119 33 U.S.C. § 1251(e).

n120 See 40 C.F.R. §§ 124.51-124.66 (2004).

n121 *Id.*

n122 *Id.*

n123 See EPA Draft Strategy to Control Nonpoint Source Pollution.

n124 *Id.*

n125 See PERCIVAL, *supra* note 21, at 491-94.

n126 *Id.* at 492.

n127 42 U.S.C. §§ 7401-7671 (2004).

n128 See PERCIVAL, *supra* note 21, at 494.

n129 42 U.S.C. § 7413(g) (2004).

n130 42 U.S.C. § 7427(a) (2004).

n131 42 U.S.C. § 7607(d), (h) (2004).

n132 See 40 C.F.R. §§ 51.102, 51.285(2004).

n133 See generally Janet S. Hathaway, The Clean Air Act and the Intermodal Surface Transportation and Efficiency Act, in AMERICAN BAR ASSOCIATION, PUBLIC PARTICIPATION 28, 31-32 (1994) (annual conference report of the Public Services Division, Standing Committee on Environmental Law of the ABA); Spyke, *supra* note 1.

n134 See UNITED STATES ENVIRONMENTAL PROTECTION AGENCY, EPA LAKEWALK MANUAL: A GUIDEBOOK FOR CITIZEN PARTICIPATION (1996); Spyke, *supra* note 1, at 283.

n135 Spyke, *supra* note 1, at 283.

n136 *Id.*

n137 See Hathaway, *supra* note 133, at 28-29.

n138 *Id.*

n139 *Id.*

n140 33 U.S.C. § 1365; 42 U.S.C. § 7604; see Mullikin & Smith, *supra* note 7.

n141 See generally JEFFREY G. MILLER, CITIZEN SUITS: PRIVATE ENFORCEMENT OF FEDERAL POLLUTION CONTROL LAWS (1987).

n142 See Eileen Gauna, Federal Environmental Citizen Provisions: Obstacles and Incentives on the Road to Environmental Justice, 22 ECOLOGY L.Q. 1, 40-1 (1995).

n143 *Id.*

n144 Mullikin & Smith, *supra* note 7, at 76.

n145 42 U.S.C. §§ 9601-9675 (2004).

n146 CERCLA was nearly a direct response to the Love Canal disaster in Niagara, New York in 1978. For an excellent treatment of this event and the tremendous public response it triggered, see JUDITH A. LAYZER, THE ENVIRONMENTAL CASE: TRANSLATING VALUES INTO POLICY 52-77 (2002).

n147 See PERCIVAL, *supra* note 21, at 224.

n148 42 U.S.C. §§ 9604, 9606 (2004).

n149 42 U.S.C. § 9607 (2004).

n150 42 U.S.C. § 9613(a) (2004).

n151 42 U.S.C. § 9613(k) (2004) (removal actions deal with emergency removal while remedial actions tend to be long-term clean-up projects that are more permanent in effect).

n152 According to CERCLA and the APA, a reviewing administrative court may only look to the administrative record. 42 U.S.C. § 9613(k) (2004); 5 U.S.C. § 557 (2004).

n153 42 U.S.C. § 9613(k)(2)(B) (2004).

n154 *Id.*

n155 42 U.S.C. § 9617(e) (2004).

n156 42 U.S.C. § 9622(d)(2) (2004).

n157 42 U.S.C. § 9659(a)(2) (2004).

n158 See Clinton County Comm'rs v. EPA, 116 F.3d 1018, 1024 (3d Cir. 1997); Schalk v. Reilly, 900 F.2d 1091, 1095 (7th Cir. 1990).

n159 See Spyke, *supra* note 1, at 287 (citing Lawrence E. Susskind, Overview of Developments in Public Participation, in AMERICAN BAR ASSOCIATION, PUBLIC PARTICIPATION 2, 2 (1994) (annual conference report of the Public Services Division, Standing Committee on Environmental Law of the ABA)).

n160 *Id.*

n161 See Lawrence E. Susskind, Overview of Developments in Public Participation, in AMERICAN BAR ASSOCIATION, PUBLIC PARTICIPATION 2, 4 (1994) (annual conference report of the Public Services Division, Standing Committee on Environmental Law of the ABA).

n162 See EPA Green Communities -- Who We Are, Environmental Protection Agency, at http://www.epa.gov/ greenkit/whoweare.htm (last modified Nov. 10, 2003) [hereinafter Green Communities Program].

n163 See Spyke, *supra* note 1, at 288.

n164 See Green Communities Program, *supra* note 162.

n165 *Id.*

n166 See Exec. Order No. 12,898, 59 Fed. Reg. 7629 (Feb. 11, 1994).

N167 See Spyke, *supra* note 1, at 290; see also Douglas McWilliams, Environmental Justice and Industrial Redevelopment: Economics and Equality in Urban Revitalization, 21 ECOLOGY L.Q. 705, 724 (1994).

n168 Mullikin & Smith, *supra* note 7, at 90.

n169 See, e.g., Rena I. Steinzor, Regulatory Reinvention and Project XL: Does the Emperor Have Any Clothes?, 26 ENVTL. L. REP. 10,527 passim (Oct. 1996); W.R. Derrick Sewell & Timothy O'Riordan, The Culture of Participation in Environmental Decisionmaking, 16 NAT. RESOURCES J. 1, 17 (1976); Daniel A. Farber, Environmentalism, Economics, and the Public Interest, 41 STAN. L. REV. 1021, 1034 (1989); Susskind, *supra* note 161.

n170 Farber, *supra* note 169.

n171 See Spyke, *supra* note 1, at 273.

n172 *Id.*

n173 *Id.*

n174 *Id.* at 274.

n175 *Id.* at 275.

n176 See Aarhus Convention, *supra* note 9.

n177 See Sean T. McAllister, The Convention on Access to Information, Public Participation in Decision-Making, and Access to Justice in Environmental Matters, 1998 COLO. J. INT'L ENVTL. L. & POL'Y 187 (1998).

n178 See UNITED NATIONS ECONOMIC COMMISSION FOR EUROPE, Aarhus Convention, available at http://www.unece.org/env/pp/ (last visited Feb. 5, 2005).

n179 *Id.*

n180 McAllister, *supra* note 177, at 189. Aarhus Convention, *supra* note 9, arts. 5(1)(b), 6(1)(a), 6(11), 9(1), and Annex I.

n181 See *infra* note 203 and accompanying text.

n182 McAllister, *supra* note 177, at 192.

n183 Namely former Eastern bloc countries and Turkey.

n184 McAllister, *supra* note 177, at 192.

n185 See *infra* note 203 and accompanying text.

n186 Popovic, *supra* note 2, at 691.

n187 Here, international initiatives are understood to mean conventions, agreements, declarations, general policy statements, and other forms of non-binding and non-enforceable international policy instruments.

n188 See Somrudee Nicro et al., Public Involvement in Environmental Issues: Legislation, Initiatives and Practice in Asian Members of ASEM Countries, in URBANIZATION AND ENVIRONMENT PROGRAM, PUBLIC INVOLVEMENT IN ENVIRONMENTAL ISSUES IN THE ASEM -- BACKGROUND AND OVERVIEW 31-43 (Thailand Environment Institute & Finnish Environment Institute) (2002) [hereinafter ASEM Report]. The report was disseminated at the fourth annual ASEM Meeting in Copenhagen, Denmark, Sept. 22-24, 2002.

n189 G.A. Res. 217A, U.N. GAOR, 3d Sess., at 139, U.N. Doc. A/810 (1948).

n190 *Id.* art. 19.

n191 G.A. Res. 2200, U.N. GAOR, 21st Sess., Supp. No. 16, at 55, U.N. Doc. A/6316 (1966).

n192 Popovic, *supra* note 2, at 686. (The early focus, of course, was on the emerging post-colonial states, and the United Nations believed that public participation would be necessary for successful development).

n193 See ASEM Report, *supra* note 188.

n194 See *infra* notes 200-26 and accompanying text.

n195 See ASEM Report, *supra* note 188.

n196 Report of the United Nations Conference on the Human Environment, at 3, U.N. Doc. A/CONF.48/14 (1972), revised by U.N. Doc. A/CONF.48/14/Corr. I (1972) [hereinafter Stockholm Declaration 1972].

n197 See ASEM Report, *supra* note 188.

n198 Stockholm Declaration 1972, *supra* note 196.

n199 *Id.*

n200 Popovic, *supra* note 2, at 700.

n201 See WORLD COMMISSION ON ENVIRONMENT AND DEVELOPMENT, OUR COMMON FUTURE, (1987) [hereinafter Brundtland Report] (defining sustainable development as development that meets present needs without compromising the ability of future generations to meet their own needs).

n202 *Id.*

n203 *Id.*

n204 United Nations Conference on Environment and Development, Rio Declaration on Environment and Development, U.N. Doc. A/CONF.151/5/Rev.l (1992) [hereinafter Rio Declaration].

n205 Popovic, *supra* note 2, at 700.

n206 *Id.*

n207 *Id.* at 701.

n208 See ASEM Report, *supra* note 188.

n209 *Id.*

n210 World Charter for Nature, G.A. Res. 37/7, U.N. GAOR, 37th Sess., Supp. No. 51, at 18, U.N. Doc. A/37/51 (1982) (referencing principle 23).

n211 See, e.g., Draft International Covenant on Environment and Development, art. 12, U.N. Doc. A/CONF.151/PC/WG.III/4 (1991); Bruntland Report, *supra* note 201.

n212 See Rio Declaration, *supra* note 204.

n213 *Id.* at 878.

n214 See *supra* note 2 and accompanying text.

n215 Popovic, *supra* note 2, at 699.

n216 See generally, e.g., United Nations Conference on Environment and Development, Agenda 21, U.N. Doc. A/CONF.151/PC/100/Add. 1 (1992) [hereinafter Agenda 21]; Malmo Declaration, U.N. UNEP, 6th Sess., Doc. UNEP/GC/SS.VI/1 (2000).

n217 See Popovic, *supra* note 2.

n218 See Agenda 21, *supra* note 216, at 219.

n219 See ASEM Report, *supra* note 188.

n220 *Id.*

n221 *Id.*

n222 Letter of the Conference on Environment and Development, 46th Sess., Agenda Items 34, 77(e)-(h), 78 & 79, at 4, U.N. Doc. A/46/632 (1991) [hereinafter Arab Declaration].

n223 See Popovic, *supra* note 2, at 688.

n224 See Arab Declaration, *supra* note 222.

n225 *Id.*

n226 Inter-American Program of Action for Environmental Protection, Resolution of the General Assembly of the Organization of American States (OAS), AG/RES. 1114 para. (g), (u) (1991).

n227 *Id.*

n228 This example is derived directly from one of the authors' experiences in Cambodia as corporate counsel for an involved party in this matter. The author had the unique experience of witnessing the public angst that can result in unmitigated environmental degradation in a country which lacks proper participatory structures.

n229 See generally Alan K.J. Tan, APCEL Report: Cambodia, Preliminary Assessment of Cambodia's Environmental Law, Asia Pacific Center for Environmental Law, available at http://sunsite.nus.edu.sg/apcel/ dbase/cambodia/reportc.html (last visited Feb. 5, 2005). For instance, while Cambodia has a general environmental law on the books allowing for the prosecution of individuals found responsible for destroying the environment or endangering public health, there is currently: (1) little oversight or regulation regarding the disposal of hazardous waste, (2) little or no public say in how hazardous waste is handled within Cambodia, (3) no knowledge of whether, when, how, or whose waste is being disposed of, and (4) no control of the process for doing so. *Id.*

n230 See Seth Mydans, Cambodia Town's 'Luck' Leaves Illness in Its Wake, N.Y. TIMES, Jan. 4, 1999, at A3; Cambodia Sends Toxic Waste Back Home, BBC News (Apr. 2, 1999), at http://news.bbc.co.uk/l/hi/world/asiapacific/ 310362.stm (last visited Feb. 5, 2005).

n231 See Cambodia Sends Toxic Waste Back Home, *supra* note 230.

n232 *Id.*

n233 See Human Rights Watch, Human Rights, Justice and Toxic Waste in Cambodia, at http://www.hrw.org/ reports/1999/cambotox (last visited Feb. 5, 2005).

n234 *Id.*

n235 *Id.*

n236 *Id.*

n237 See Danielle Knight, Toxic Waste Heads Back to Taiwan, ASIAN TIMES ONLINE (Apr. 3, 1999), at http://www.atimes.com/china/AD03Ad01.html (last visited Feb. 5, 2005).

n238 See *id.*; see also Marc Lifsher, Toxic Waste from Asia Head to State, WALL ST. J., Mar. 24, 1999, at CA1.

n239 See Knight, *supra* note 237.

n240 See U.S. AGENCY FOR INT'L DEV., CAMBODIA (1999) at http://www.usaid.gov/pubs/cp99/ane/kh.htm (last visited Feb. 5, 2005); INT'L INST. FOR ENV'T AND DEV., REPORT OF THE CONFERENCE ON THE ROLE OF PUBLIC PARTICIPATION 16 (2001), at http://www.iied.org/mmsd/mmsd_pdfs/IBA_public_participation.pdf (last visited Feb. 5, 2005); UNITED NATIONS, ECONOMIC AND SOCIAL COUNCIL, COMMISSION ON HUMAN RIGHTS, ECONOMIC SOCIAL AND CULTURAL RIGHTS, COMMISSION ON HUMAN RIGHTS para. 61-66, U.N. Doc. E/CN.4/2000/50 (2000).

n241 Popovic, *supra*, note 2 at 708.

n242 See Edward McCutcheon, Think Globally, Act Locally: Promoting Effective National Environmental Regulatory Infrastructures in Developing Nations, 31 CORNELL INT'L L.J. 395 (1998).

n243 *Id.*

n244 See THE WORLD BANK, World Development Report: Infrastructure for Development (1994), available at http://www.worldbank.org/archives/dev_94023.htm (last visited Feb. 5, 2005).

n245 McCutcheon, *supra* note 242, at 403-6 (citing nine categories of environmental degradation: deteriorating air quality, deteriorating water quality, atmospheric change, environmental deterioration in urban areas as result of population growth, soil degradation and deforestation, destruction of biodiversity, deterioration of coastal resources, pollution from agro-chemicals, and deterioration of natural and cultural heritage).

n246 *Id.* at 397.

n247 *Id.*

n248 *Id.*

n249 *Id.*

n250 See *supra* notes 227-37 and accompanying text.

n251 See *supra* Parts III & IV.

n252 Tran Tuan Anh, Vietnam's Co-operation for Development in Mekong River Basin, NHAN DAN, March 13-14, 2004, available at http://www.nhandan.org.vn/ english/20040313/bai-eco2.html (last visited Feb. 5, 2005).

n253 *Id.*

n254 *Id.*

n255 Tun Myint, Globalization, Democracy and Domestic Law: Democracy in Global Environmental Governance: Issues, Interests, and Actors in the Mekong and the Rhine, 10 IND. J. GLOBAL. LEG. STUD. 287 (2003).

n256 *Id.*

n257 See *id.* at 300.

n258 *Id.* at 299.

n259 *Id.*

n260 *Id.* at 300.

n261 Under the plan, the MRC will cooperate to seek investment for development plans designed to build sustainable practices in irrigated agriculture, watershed management, fisheries, hydropower, navigation, transport, water-related tourism and flood control. See MEKONG RIVER COMMISSION FOR SUSTAINABLE DEVELOPMENT, BASIN DEVELOPMENT PLAN, at http://www.mrcmekong.org/programmes/bdp/bdp.htm (last visited Feb. 5, 2005).

n262 *Id.*

n263 See MEKONG RIVER COMMISSION FOR SUSTAINABLE DEVELOPMENT, *supra* note 261.

n264 *Id.*

n265 See Myint, *supra* note 255.

Appendix B

Community Participation in Environmental Protection

Thomas Stowe Mullikin*, Nancy S. Smith**

UCLA Journal of Environmental Law and Policy, Volume 21, Number 75 (2002/2003)

Reprinted with Permission of the Publisher, UCLA Journal of Environmental Law and Policy © 2003.

BIO:

* J.D., 1986, University of South Carolina School of Law; B.A., 1982 cum laude, Phi Beta Kappa, University of South Carolina. Mr. Mullikin is a member of Moore & Van Allen PLLC and practices in the firm's Charlotte office. As leader of the firm's corporate environmental team, his practice focuses on the intersection among governmental, environmental and business interests.

** J.D., 1998, University of North Carolina at Chapel Hill School of Law, Member of the Holderness Moot Court Negotiations Team; B.A., 1991, University of North Carolina at Chapel Hill. Mrs. Smith is an associate of Moore & Van Allen PLLC in the Charlotte office where she practices with the firm's corporate and environmental groups.

SUMMARY:
... Much has been written in recent years concerning the erosion of citizens' rights to file suits alleging environmental harm against industry. ... This model provides a progressive alternative to citizen suit litigation by promoting a partnership between a community and an industry that can effectively address and rectify a community's environmental concerns. ... Such collaborative and innovative environmental protection efforts between a state agency and regulated industry may be thwarted by the

threat of potential citizen suits. ... The requisite analysis for standing focuses on the plaintiff rather than the environment despite any alleged injury to the environment that the plaintiff is being accused of causing. ... Laidlaw cross-appealed both claiming that the citizens lacked standing because they had suffered no injury in fact and that the suit should have been barred since DHEC had diligently prosecuted a prior action regarding the same violations. ... However, relying on the decision in Steel Co., the court of appeals held that, because any civil penalty would be paid to the U.S. Treasury, such penalties could not redress any injury suffered by a citizen plaintiff. ... Rather, it must be the result of a real and perceived partnership among individuals who have the authority to negotiate and act on behalf of their respective stakeholder groups.

TEXT:
[*75]

I.   Introduction

Much has been written in recent years concerning the erosion of citizens' rights to file suits alleging environmental harm against industry. An article in Bergen, New Jersey's Record newspaper typifies these writings: "You don't need a chainsaw to damage the environment. A gavel and a black robe can be far more effective." 1 Of course, commentary such as this tends to shed more heat than light on the subject by condemning integral decision-makers and shifting focus away from substantive issues. The purpose of this article, however, is to approach the issue in a manner that illuminates the role of citizen participation in environmental protection rather than descending into argument. Analyzed herein are the two primary vehicles which citizens have successfully used to seek redress for alleged environmental harm: (1) citizen suits under the major environmental bodies of law; and (2) civil rights Title VI actions for environmental justice. Finally, the Collaborative Compact Model is proposed and analyzed. This model provides a progressive alternative to citizen suit litigation by promoting a partnership between a community and an [*76] industry that can effectively address and rectify a community's environmental concerns.

II. Origins of Citizen Suit Provisions

One of the basic features of our nation's environmental protection system is the right of citizens to sue in federal court to force industry to comply with environmental standards. Congress included citizen suit provisions in major environmental laws, such as the Clean Water Act 2 ("CWA") and the Clean Air Act, 3 to enable citizens to act as "private attorneys general" and to supplement the government's limited enforcement resources. 4 These provisions ensure the rigorous enforcement of environmental laws when states do not or will not address serious pollution problems in good faith. Citizen suits have been a narrow, but very important, statutory remedy.

It is important to note that these provisions authorize citizens to complement government action, not to compete with or replace it. For example, section 505 of the CWA authorizes any citizen to commence a civil action against a person or entity "alleged to be in violation" of an effluent standard or limitation. 5 The citizen-plaintiff must provide notice of the alleged violation to the Administrator of the U.S. Environmental Protection Agency, to the enforcement agency of the state where the alleged violation occurs, and to the alleged violator at least sixty days before he may file suit. 6 The 60-day notice requirement allows for the industry to come into compliance and provides the government with an opportunity to act in lieu of the citizen-plaintiff to avoid multiple lawsuits.

As indicated in the Senate Report on the Federal Water Pollution Control Act Amendments of 1972 7 ("FWPCA"), "the Committee intends the great volume of enforcement actions to be brought by the State" and that citizen suits are proper only "if [*77] the Federal, State, and local agencies fail to exercise their enforcement responsibility." 8 Further, citizen suits are to "ignite agency enforcement" and to act as an "alternative enforcement mechanism absent agency enforcement." 9 Thus, a citizen suit is not authorized unless the state has failed or declined to take action or has not "diligently prosecuted" a claim after receipt of the 60-day notice. 10

Congress' aversion to the duplicative actions that can result from simultaneous or subsequent citizen suits is also illustrated by the "diligent prosecution" provision of the CWA. 11 This provision mandates that no citizen suit may be undertaken "if the Administrator or State has commenced and is diligently prosecuting a civil or criminal action in a court of the United States, or a State to require compliance with the standard, limitation, or order." The meaning of these provisions could not be clearer—citizen suits are barred when the state has commenced and is diligently prosecuting a civil action in court to require compliance. Further, the relief authorized under the CWA includes injunctive relief and penalties that are payable to the U.S. Treasury, not to the private citizens. This eliminates the financial incentive that might otherwise encourage competitive litigation and supports the premise that "private attorneys general" actions are meant to complement government action, rather than to compete with or replace it.

Further, courts have expressed a clear aversion to multiple suits. "Duplicative actions aimed at exacting financial penalties in the name of environmental protection at a time when remedial measures are well underway do not further [our] goal. They are, in fact, impediments to environmental remedy efforts." 12 Similarly, the U.S. Supreme Court has often described the role of citizen suits as supportive and has recognized that state agencies administering the FWPCA are to be given significant discretion. The Court noted in one case that the bar on citizen suits during [*78] governmental enforcement action clearly indicates "that the citizen suit is meant to supplement rather than supplant governmental action." 13

III.   Key Question: Did the State Diligently Prosecute?

The threshold issue in determining whether a citizen suit should be barred is whether the state "diligently prosecuted" the subject industry. An analysis of whether diligent prosecution has taken place neither encompasses de novo assessment of the agency's procedure in seeking enforcement and computing sanctions nor requires a comparison of the agency's activities with what the citizen group or the court may prefer. In fact, a heavy burden has been placed on citizens who allege that the state has not diligently prosecuted: "The court must presume the diligence of the State's prosecution of a defendant absent persuasive evidence that the State has engaged in a pattern of conduct in its prosecution of the defendant that could be considered dilatory, collusive or otherwise in bad faith." 14 Instead, to overcome the presumption that the state agency did prosecute diligently, citizen-plaintiffs must show that the actions by the state agency amounted to inaction. 15

Diligence is the act of remedying the violations in any manner that the state decides. The mere fact that the state does not take the precise action that plaintiffs would prefer does not constitute lack of diligence. 16 Therefore, absent fraud or collusion, a citizen suit that visits identical issues in order to seek a remedy other than the state sought is inappropriate. While a citizen may be dissatisfied with an agency's ultimate resolution, the filing of a citizen suit out of dissatisfaction would impinge on federal and state agency authority and autonomy and create the possibility for duplicative actions. For example, the fact that the state "alleged fewer separate violations" and that it "sought a less substantial civil penalty" was not enough to overcome the presumption of "diligent prosecution." 17

[*79] The Act authorizes injunctions and civil penalties which are payable to the U.S. Treasury. 18 In determining the amount of a civil penalty, the court must evaluate "the seriousness of the violation ... the economic benefit (if any) resulting from the violation, any history of such violations, any good-faith efforts to comply with the applicable requirements, the economic impact of the penalty on the violator, and such other matters as justice may require." 19 However, the First Circuit has held that diligent prosecution by a state agency does not require assessing any penalty at all. In North and South Rivers Watershed Assn. vs. Scituate, 20 the court explained that a decision otherwise would expand the "supplemental" role envisioned for such suits and would create a potentially intrusive role for citizens. 21 A citizen suit may not be brought merely to obtain attorneys' fees incurred in pursuit of a citizen suit when the agency has already remedied the problem causing the violations. 22 Since citizens suing under the FWPCA are deemed "private attorneys general," there is little left to do after the government has negotiated an agreement. 23

IV.   State Enforcement and Settlements -- Water Pollution Example

States are afforded the opportunity to establish their own water quality standards and discharge limits in individual permits they issue. 24 Many states have adopted more stringent standards than the federal limits require in an effort to achieve lower levels of pollution in their waters. 25 A number of states also encourage the development of innovative technologies to achieve lower levels of pollution. Such decisions may affect the method of enforcement [*80] by the state agencies. For example, a state agency may require that a company expend money for new equipment or implementation of new technology in lieu of a strict penalty. 26 In addition, states are afforded latitude in selecting the specific mechanisms of their enforcement program and are given great deference to proceed in a manner they consider to be in the best interests of all parties involved.

This shift in enforcement emphasis from federal to state agencies is clearly reflected in the 1987 amendments to the FWPCA. 27 The FWCPA created a program "to restore and maintain the chemical, physical, and biological integrity of the nation's waters." 28 In addition, the FWCPA recognizes Congress's vision that states would monitor the performance of their permitted industries and take administrative, civil and criminal enforcement action in response to violations as necessary to protect the quality of the waters within their jurisdictions. The extent of enforcement employed is often dependent upon the specific circumstances of each violation. Most violations are resolved through negotiations and the issuance of administrative and judicially approved consent agreements and orders. The orders provide for a compliance timetable, the payment of civil penalties or the performance of supplemental environmental projects in lieu of some portion of the civil penalties. Negotiations are conducted between the regulator and the regulated industry whereby an agency may be willing to forego a higher penalty in order to obtain greater protection of the water quality. As part of its enforcement discretion, the agency may require monitoring and prevention of future contamination, including requiring the installation or implementation of new technology or equipment. 29

Such collaborative and innovative environmental protection efforts between a state agency and regulated industry may be thwarted by the threat of potential citizen suits. In essence, these suits would second-guess an agency's discretion after the settlement. In cases where the plaintiff/citizen groups have attempted to impose their judgment after an agency has negotiated to protect [*81] waterways and force an extreme penalty through judicial fiat, courts have correctly denied them access to this remedy. It is critical that the agencies and their industrial and municipal permittees have confidence in the enforcement process and the level of resolution that it achieves. When that process is undertaken in good faith, the parties should not face lingering concerns over whether there may be a citizen group or a federal court that regards the settlement as less than the required "diligent prosecution." 30 If industry is subject to additional enforcement action, namely the prospect of penalties in excess of those already imposed by the agency, then they are less likely to negotiate the resolution of their violations. Clearly, this would result in the unnecessary proliferation of litigation. Even state courts would not be safe from scrutiny as to whether the penalties they imposed or approved are strict enough and whether the injunctive relief they awarded or approved is sufficient to be considered a "diligent prosecution."

In short, allowing a citizen suit to proceed in such circumstances has the very real potential of seriously disrupting the enforcement discretion that is necessarily inherent in the delegation of the enforcement of federal environmental programs to the states. The Supreme Court has also recognized the inherent danger in permitting citizen suits to revisit the efficacy of governmental enforcement activities creating a disincentive to settle. In Gwaltney of Smithfield Ltd. v. Chesapeake Bay Foundation, 31 the Court considered the following hypothetical:

> Suppose that the Administrator [of the EPA] identi-
> fied a violator of the Act and issued a compliance or-
> der under 309(a). Suppose further that the Adminis-
> trator agreed not to assess or otherwise seek civil pen-
> alties on the condition that the violator take some ex-
> treme corrective action, such as to install particularly

effective but expensive machinery that it otherwise would not be obliged to take. If citizens could file suit, months or years later, in order to seek the civil penalties that the Administrator chose to forego, then the Administrator's discretion to enforce the Act in the public interest would be curtailed considerably. The same might be said of the discretion of state enforcement authorities. Respondents' interpretation of the scope of the citizen suit [i.e., that such suits should be permitted to recover civil penalties for past violation] would change the nature of the citizens' role from interstitial to [*82] potentially intrusive. We cannot agree that Congress intended such a result. 32

Duplicative actions aimed at exacting financial penalties in the name of environmental protection do not further the goal of environmental laws. States should maintain the flexibility to develop strong, innovative policies and to cooperate with regulated industry. This flexibility will allow them to continue to encourage technological advances and processes to achieve the federal objective of restoring and maintaining the integrity of the environment. Such flexibility recognizes that problem solving and collaboration are at least as effective in achieving these national goals as enforcement through litigation.

V.   The High Court Responds

A quick review of pertinent Supreme Court citizen suit jurisprudence draws attention to three cases. First, in Gwaltney, citizen-plaintiffs alleged past and ongoing violations of the Clean Water Act. 33 The court of appeals affirmed the lower court ruling that citizens could bring action for past violations alone. 34 However, the Court disagreed, holding that an action could not be brought based only on past violations. The Court remanded the case to determine if there was a good faith allegation of ongoing violations. 35 In its ruling, the Court discussed at length the role of citizen suits and emphasized that they are meant to "supplement, not supplant" governmental action. 36 In fact, the legislative history of the Act clearly indicates that the roles of a citizen-plaintiff are to enforce a current violation and to halt ongoing pollution. 37 The Court further stressed that:

> if we assume, as respondents urge, that citizen suits may target wholly past violations, the requirements of notice to the alleged violator becomes gratuitous. Indeed, respondents, in propounding [*83] their interpretation of the Act, can think of no reason for Congress to require such notice other than "it seemed right" to inform an alleged violator that it was about to be sued.

Second, in Steel Co. v. Citizens for a Better Environment, 38 the citizens filed a private enforcement action for declaratory and injunctive relief alleging that Steel Co. had failed to file timely reports required by the Emergency Planning and Community Right-To-Know Act of 1986 39 ("EPCRA"). The district court first found that, because the filings were brought up-to-date by the time the complaint was filed, it lacked jurisdiction to hear the suit for a present violation. 40 It then found that, since EPCRA does not allow for purely historical violations, there was no claim upon which relief could be granted. The Seventh Circuit reversed the district court decision concluding that EPCRA allows citizen suits for past violations. 41 The Supreme Court reversed on the issue of Article III standing, holding that the relief sought would not remedy the citizens' alleged injury in fact. 42 Therefore, the Court determined the citizens lacked standing to maintain the suit and the lower courts lacked jurisdiction to hear it. 43 The "'irreducible constitutional minimum of standing' contains three requirements." 44 As the Court stated, there must be injury in fact, causation, and redressability.

First and foremost, there must be alleged (and ultimately proven) an "injury in fact"—a harm suffered by the plaintiff that is "concrete" and "actual or imminent, not 'conjectural' or 'hypothetical.'" Second, there must be causation—a fairly traceable connection between the plaintiff's injury and the complained-of conduct of the defendant. And third, there must be redressability—a likelihood that the requested relief will redress the alleged injury. This triad of injury in fact, causation, and redressability comprises the core of Article III's case-or-controversy requirement, and the party invoking federal jurisdiction bears the burden of establishing its existence. 45

The Supreme Court found the citizens' complaint failed the redressability test because the relief sought would not remedy [*84] injuries suffered by the plaintiffs. 46 That relief sought included: a declaratory judgment that Steel Co. had violated EPCRA; injunctive relief authorizing the citizens to make periodic inspections of Steel Co.'s facility and records; the requirement that Steel Co. give the citizens copies of its compliance reports; and, orders requiring Steel Co. to pay civil penalties and litigation costs to the U.S. Treasury. 47

The Supreme Court made clear and compelling arguments in both Gwaltney and Steel Co. to the following effect. First, citizens bringing a private enforcement action must follow the statutory standards. Second, citizens bringing a private enforcement action are no different than all other plaintiffs bringing an action and must meet Article III standing. Simply because the action is brought in the name of the environment does not mean that the plaintiffs receive special treatment from the Court. The requisite analysis for standing focuses on the plaintiff rather than the environment despite any alleged injury to the environment that the plaintiff is being accused of causing. 48 The Supreme Court indicated its distaste for a different standard of review for cases which involve environmental harm and restated prior holdings that environmental plaintiffs meet the Article III standing requirement when they "aver that they use the affected area." 49

In 2000, the Supreme Court handed down Friends of the Earth, Inc. v. Laidlaw Environmental Services (TOC), Inc. 50 This ruling may significantly impact the future of all citizen lawsuits. Laidlaw owned and operated a hazardous waste incinerator in Roebuck, South Carolina. 51 As part of that facility, Laidlaw maintained a wastewater treatment plant for water used in its air pollution control system. Laidlaw discharged the treated wastewater into the North Tyger River pursuant to a National Pollutant Discharge Elimination System ("NPDES") permit issued by the South Carolina Department of Health and Environmental Control ("DHEC"). Once it received its permit, Laidlaw repeatedly discharged mercury in excess of the permitted levels, among other violations.

[*85] It is important to note that DHEC set the mercury discharge limit at 1.3 parts per billion (ppb) for Laidlaw from 1988 through 1993, while

the limit for other facilities was typically 2.0 ppb. 52 The 1.3 ppb standard actually required Laidlaw to discharge water that was cleaner than drinking water, which has a mercury standard of 2.0 ppb. The state's initial permit limit required Laidlaw's discharge at the pipe to contain 25% less mercury than is permitted in drinking water, even before being diluted by the river.

DHEC brought a formal action against Laidlaw alleging non-compliance with the NPDES permit. 53 A consent agreement was reached and approved by the state Circuit Court. The agreement included a civil penalty of $ 100,000. It also required that Laidlaw make "every effort" to comply with the permit. Laidlaw continued to explore additional technologies and ultimately found a solution to meet the 1.3 ppb limit through experimentation utilizing activated carbon, micro-filtration, and ion exchange.

Immediately after the DHEC action, the Friends of the Earth ("FOE") and Citizens Local Environmental Action Network, Inc. ("CLEAN") brought an action against Laidlaw in U.S. District Court alleging noncompliance with the NPDES permit. The plaintiffs sought declaratory and injunctive relief, the imposition of civil penalties, and the award of costs including attorneys' and witness fees. The lower court found that the plaintiffs had Article III standing to bring the suit. 54

The lower court denied Laidlaw's motion to dismiss on the grounds that the citizen suit was barred under Section 505(b)(1)(B) of the CWA. 55 The court held that DHEC had not "diligently prosecuted" the action because the agency did not calculate and review the economic benefit Laidlaw received by continually failing to comply with the permit even after the lawsuit was initiated. 56 The court also found that there was no fraud or collusion between DHEC and Laidlaw and specifically ruled that Laidlaw had acted in good faith in working with the state enforcement agency. 57 It found that no harm had occurred to the environment, that Laidlaw had expended in excess of $ 1 million [*86] in various attempts to achieve compliance, and that compliance had in fact been achieved. The district court ultimately denied the plaintiff's equitable relief, but did impose a penalty of $ 405,800. 58 This sum was deemed adequate after taking into account the "total deterrent effect" of the judgment and that Laidlaw was required to pay plaintiff's legal fees. 59

The citizens appealed the adequacy of the penalty. Laidlaw cross-appealed both claiming that the citizens lacked standing because they had suffered no injury in fact and that the suit should have been barred since DHEC had diligently prosecuted a prior action regarding the same violations. 60 The Fourth Circuit dismissed the action on the grounds that it had become moot. 61 In doing so, the court emphasized that the three elements of standing must continue to exist at every stage of review, not merely at the time of the filing of the complaint. 62 Since FOE did not appeal the denial of injunctive or declaratory relief, the only potential relief available to redress their claimed injuries would be the civil penalty. However, relying on the decision in Steel Co., the court of appeals held that, because any civil penalty would be paid to the U.S. Treasury, such penalties could not redress any injury suffered by a citizen plaintiff. 63 The court of appeals concluded that a citizen suitor's claim for civil penalties must be dismissed as moot because the only remedies available upon appeal were civil penalties payable to the government, and therefore the damages assessed would not redress the injury to the plaintiff. 64

In January 2000, Justice Ginsburg, writing for the Court, reversed the appellate court's decision. 65 Justice Ginsburg opined for the majority:

> In directing dismissal of the suit on ground of mootness, the Court of Appeals incorrectly conflated our case law on initial standing to bring suit ... with our case law on post-commencement mootness ... . A defendant's voluntary cessation of allegedly unlawful conduct ordinarily does not suffice to moot a case. The Court of Appeals also misperceived the remedial potential of civil penalties. Such penalties may serve, as an alternative to an injunction, [*87] to deter future violations and thereby redress the injuries that prompted a citizen suitor to commence litigation.

The Court also distinguished this case from Steel Co., which did not reach the issue of standing to "seek penalties for violations that are on-going at the time of the complaint and that could continue into the future." 66 Thus, the case was remanded for further consideration on the issue of mootness and payment of attorney's fees. 67

Looking beyond the scholarly legal opinions, Laidlaw raises several important questions. First, did the state regulators do their job and was enforcement rigorous? Yes. A DHEC agent was on-site at the facility on a daily basis. DHEC closely monitored the feasibility studies and established schedules for the installation and implementation of new technologies. Penalties were voluntarily assessed against Laidlaw on two different occasions, one totaling $ 20,000, the other, $ 100,000. Second, was there harm to the environment? No. After numerous days of testimony, the District Court specifically found that no harm to the environment had occurred. 68 Third, was anything achieved by this lawsuit beyond that accomplished by the settlement agreement? No. Compliance was achieved by the industry through its own efforts in cooperation with the regulating agency. The courts, state, industry and environmentalists all spent untold hours pursuing or defending a lawsuit over seven years that resulted in protections that were no more beneficial to the environment than the terms of the consent order. These insights illustrate how traditional litigation often fails to achieve its intended goal of protecting the environment. Instead, seeking redress through the court system often costs money and expends valuable time with a result indistinguishable from an environmental perspective as the alternative of cooperation.

VI.   Environmental Justice

Unfortunately, Friends of the Earth is all too typical of most environmental debates. It pitted industry against activists, rhetoric against reality, and posturing against partnering. Laidlaw never stated it believed there was no value to citizen suits. Instead, [*88] the company repeatedly stated its position that this case should be dismissed since DHEC's actions qualified as diligent prosecution precluding litigation. 69

Legal analysis aside, the larger question at issue is whether there are ways for citizens and industries to address their concerns without resorting to lawsuits. The answer can be found in an area that is just as contentious as citizen lawsuits, if not more so—environmental justice. Like citizen suits, citizen activity has also increased in the area of environmental justice as plaintiffs seek "environmental equity" through Title VI of the Civil Rights Act. 70

The Environmental Justice Movement began in 1982 when hundreds of African-Americans objected to North Carolina's plan to bury PCBs in a proposed landfill in Warren County. 71 They were joined by leaders of the United Church of Christ, the Southern Leadership Council, and Walter Fauntroy, U.S. House delegate from the District of Columbia. 72 The protestors believed that Warren County had been unjustly chosen for the landfill because it was one of the poorest regions of the state and had the highest percentage of people of color of any North Carolina county. 73 After 400 demonstrators were arrested amid a surge of national attention, Governor Jim Hunt declared that no new landfills would be built in Warren County. 74 The location of the landfill was a political decision. There is evidence that the site was not the most suitable from an environmental or scientific perspective, but it was chosen as the place to locate the landfill given the demographic composition of the community. 75

Two decades later, little consensus exists as to the validity of charges that environmental threats are being forced upon communities of color because those areas do not possess the political or economic strength to resist. However, four major studies undertaken between 1982 and 1992 do support this argument. 76 All [*89] reached a strikingly similar conclusion: exposure to hazardous wastes was falling disproportionately upon minority and/or low-income populations. 77

In a contrasting study, 78 a comprehensive examination of environmental justice across the state of South Carolina analyzed the areas surrounding 59 non-governmental facilities that had Resource Conservation and Recovery Act of 1976 79 Part B permits for the treatment, storage and disposal of hazardous wastes. It evaluated demographics in those areas according to zip codes, census tracts, and a 2.5-mile radius around each site. 80 The study found that racial compositions near these facilities were within three percent of the state average. 81 Beyond that, it also concluded that: 1) a majority of the hazardous waste facilities were located in areas where the percentage of minorities was lower than the state average; 2) the facilities were located in areas that had a smaller percentage of people below the poverty level than did the state as a whole; and, 3) there was no disparity in the way regulators enforced environmental rules in minority communities as opposed to others. 82 After almost twenty years of study, this issue is still defined more by conflict than common ground. Environmental justice remains mired in a policy war of attrition that pits citizen activists against the business community. "To reach the promise of [a] new dialogue is to escape from gridlocked, two-way discourses between traditional environmental groups and industry. Both are now locked into a rhetoric of combat and fault that places [*90] communities and their immediate concerns at a distance." 83 Rather than forging a constructive dialogue that at least respects the concerns of all stakeholders, the debate is too often fueled by agitation, self-interest, politics, partisanship, and rhetoric. The movement of all stakeholders beyond the current charge versus countercharge scenario will require a new model that acknowledges two very important realities. First, to the extent that industrial facilities are located in minority communities, the people who live there tend not to gain any significant economic advantages from those sites. Second, the needs and interests of the community and business do not necessarily conflict. If there is ever to be a constructive discourse on environmental justice, both sides must recognize this truth.

VII. Litigation Alternatives

The availability of citizen suits has given environmental advocacy groups a powerful litigation tool. However, this forum for controversial environmental decisions has limited accessibility for most environmentally assaulted people. A new model being proposed herein uses the realities stated above as a point of departure. It unites basic theories of receiver-centered communications with the win-win objectives of negotiated rule-making. It is based on the belief that there is a clear parallel between rational regulatory promulgation and reasoned environmental debate.

A.   The Collaboration Compact Model

This new paradigm, called the Collaboration Compact Model, uses a seven-step process to create a binding corporate-community partnership. The steps are as follows:

1. Assessment of company and community needs and an initial determination of where they might coincide; identification of each stakeholder's objective and priorities facilitates this evaluation;

2. Alignment of those needs to reflect the mutual interests of both primary stakeholders leads to discovery of areas of agreement;

3. Concurrence as to what constitutes a mutually acceptable outcome; creation of a "win-win" result for both parties;

[*91] 4. Development of a comprehensive program designed to secure the desired outcomes and unite the interests of each party; honest and open communication between shareholders and impacted third parties is required;

5. Modification of the initial plan either to fine-tune and/or clarify objectives and strategies or to reflect changes in circumstances or events;

6. Acceptance of the program with a finalized memorandum of understanding as to what will be expected from all parties and delineating the role of the stakeholders in assuring program success; and,

7. Implementation of a jointly approved program pursuant to the terms of a memorandum of understanding.

The successful execution of this strategy requires six characteristics from the participating stakeholders. First, there must be willingness to compromise and recognition of the value that compromise brings to the process. This model focuses on issues that unite rather than divide and maintains a positive focus to encourage completion of the program. As such, matters of singular interest to either the company or the community may remain unresolved. All parties must acknowledge and accept that priorities and goals may need to be adjusted to facilitate a successful outcome. All parties must be prepared to make concessions for the benefit of the group.

Second, there must be shared political management. This refers to the willingness and ability of those who have something at stake—the company and the community—to persuade all affected third parties of the validity of the program and the process by which it was reached. These third parties include, but are not limited to, regulators, elected officials, media, employees and members of the larger business and activist communities. This element requires that the stakeholders identify third parties early in the process and involve them in discussions to ensure cooperation. An otherwise effectively planned program can face potentially fatal roadblocks by third parties whose interests are not taken into account.

Third, there must be joint leadership. Project oversight cannot be the province of either the business or the community organization. Rather, it must be the result of a real and perceived partnership among individuals who have the authority to negotiate and act on behalf of their respective stakeholder groups. Once identified as leaders, these individuals must demonstrate a willingness [*92] and ability to work as a council focused on mutual goals rather than on unilateral interests of their respective organizations. This will ensure that when problems arise, they can be fairly and efficiently addressed without derailing the entire program.

Fourth, there should be clear-cut responsibilities. That is, the company and the community each must know exactly what they are responsible for and those responsibilities should be clearly detailed in a memorandum of understanding. This written memorial prevents misunderstandings as to each party's role and commitment to the program. It also encourages extensive input and consideration prior to implementation. Defining responsibilities clearly and in writing makes the parties accountable to each other and to the agreement.

Fifth, from the outset, there must be an agreement by both parties to adhere to the so-called "rules of the game." These rules include not only the specifics of the program such as what constitutes a "win-win," but also the guidelines for achieving program success. All rules are outlined in the memorandum of understanding and establish the basis for the relationship between the company and the community.

Lastly, there must be inclusion. In any scenario, there will be people who have the power, authority, ability or potential to impede or prevent successful completion of the program. To whatever degree possible and appropriate, these people should be brought into the process and given a role that helps assure their support. Whether private leaders or public officials, these individuals are typically prominent community members with the ability to significantly impact initiatives. Including these leaders in the development of the program can preempt possible public dissention and lends credibility to the program. In sum, by incorporating each of these points into a strategy for cooperation among the various interest-holders, a resolution suitable to all parties can be achieved. By identifying priorities to developing a cooperative program, it is possible to move away from the costly and time-consuming dispute resolution tool of litigation.

B.   The Collaboration Compact Model Applied

This model was successfully applied by a South Central Los Angeles hazardous waste treatment storage and disposal facility ("TSDF") operating in the heart of an area devastated by the [*93] Rodney King riots. 84 In mid-1998, the company began negotiations with the Concerned Citizens of South Central Los Angeles ("CCSCLA"), the state's oldest environmental justice organization. From the earliest discussions, it was clear that the group's leaders did not believe they had accrued sufficient benefits from having the world's largest recycler in their neighborhood. That became the starting point for a months-long dialogue that concentrated not on the issues that separated the company and the community, but on the interests that bound them together. Both sides agreed to reject rhetoric and use the Collaboration Compact Model to focus on realities.

The result was The Partnership for Environmental Training ("PET"). The program's objective was to improve the negative opinion in the community toward the company and to give the citizens an opportunity to benefit from the company's presence. It provided young men and women in the community with the requisite skills to seek and obtain good-wage, high-tech jobs in the environmental services industry. In 1999, a total of 72 students graduated from the program. The program included 80 hours of instruction sponsored by Safety-Kleen, Inc. at the company's Los Angeles facility. Additionally, the company underwrote the cost of pre-employment medical exams and drug screenings, paid for respirators, safety boots, work clothes, bus fares and all training materials.

A seven-member volunteer advisory council, a majority of which was appointed by the CCSCLA, managed the program. Representatives of the company occupied the three remaining seats. In addition to managing the program, the advisory council set up a temporary employment agency for each program graduate. Some were able to move into jobs with the company and others were placed in jobs with other corporations, including competitors. The goal was to ensure that the trainees had a new set of technical skills as well as the opportunity to put those skills to work in relevant, real-world situations.

PET grew from a mutual understanding of the needs and interests of both parties. Simply stated, the community needed jobs. The area's unemployment rate was approximately 52% at the time the program was implemented. Local organizations also had an interest in getting involved in programs that provided the [*94] skills the residents needed to find meaningful jobs. On the other hand, the company needed a pipeline of trained professionals and had an interest in reaching out to the community to preempt future conflicts. PET enabled both groups to secure their institutional objectives.

Most importantly, the "winners" extended beyond the primary stake-holders. Three other organizations active in the revitalization of South Central Los Angeles were invited into the decision-making process. The City of Los Angeles benefited because PET graduates have the skills necessary to assist with household hazardous waste cleanups throughout the city. By helping to rid neighborhoods of dangerous materials, they assist in creating the safest possible communities for families and children.

Other temporary employment agencies in the area also benefited. Those companies would have had to spend up to $ 1,500 per person to screen, train and otherwise prepare employees for work on the job site. Since the company agreed to absorb those costs as part of its responsibilities, the temporary agencies could provide better, more fully trained prospects at a lower internal cost. 85

Lastly, citizen-trainees clearly benefited. These individuals obtained quality job-skills training that increased their potential to acquire high paying careers in the industry. The program filled a void in the region by providing meaningful employment opportunities in an area where such opportunities were quite limited.

The Collaboration Compact Model was successfully implemented in California and resulted in benefits for all participants. The company was able to improve the community's perception of its operations and employ local residents while the host community achieved job creation and environmental improvement. Clearly, the men and women involved were provided skills training and educational opportunities that would not have been otherwise available.

VIII.   Conclusion

The conflicts between industries and their host communities are likely to continue to increase in severity until and unless innovative methods of relationship building are developed. If history is any indication, these antagonistic relationships will [*95] continue to spill over into the courts producing costly, time-consuming litigation that does not serve the environmental, community or business interests. This is especially true regarding citizen lawsuits in which community representatives feel they have been denied a voice in the process. Collaborative models, such as that defined by the Partnership for Environmental Training, have tremendous potential for building partnerships, defusing hostilities between stakeholders, and creating more win-win situations.

FOOTNOTES:

n1. Keri Powell & Robert Shapiro, U.S. Courts Unfriendly to Citizen Environmental Suits, Bergen County Rec., July 23, 1999, at L7.

n2. The Federal Water Pollution Control Act Amendments of 1972, Pub. L. No. 92-500, 86 Stat. 816 (codified as amended in scattered sections of 33 U.S.C.), are generally called the Clean Water Act.

n3. This is the common name of the Clean Air Amendments of 1970, Pub. L. No. 91-604, 84 Stat. 1676 (codified as amended in scattered sections of 42 U.S.C.).

n4. EPA v. City of Green Forest, 921 F.2d 1394, 1405 (8th Cir. 1990).

n5. 33 U.S.C. 1365(a)(1) (2000).

n6. 1365(b)(1)(A).

n7. Pub. L. No. 92-500, 86 Stat. 816 (codified as amended in scattered sections of 33 U.S.C.); see also *supra* note 4.

n8. City of Green Forest, 921 F.2d at 1403 (citing S. Rep. No. 92-414, at 64 (1971)).

n9. Conn. Fund for Env't v. Upjohn Co., 660 F. Supp. 1397, 1403 (D. Conn. 1987) (citing Baughman v. Bradford Coal Co., 592 F.2d 215, 218 (3d Cir. 1978)) (emphasis added).

n10. City of Green Forest, 921 F.2d at 1403; accord Scituate, 949 F.2d at 557; see also Conn. Coastal Fishermen's Ass'n v. Remington Arms, Co., 777 F. Supp. 173, 179 (D. Conn. 1991).

n11. 1365(b)(1)(B).

n12. N. & S. Rivers Watershed Ass'n, Inc. v. Town of Scituate, 949 F.2d 552, 556 (1st Cir. 1991).

n13. Gwaltney of Smithfield, Ltd. v. Chesapeake Bay Found., 484 U.S. 49, 60 (1987) (emphasis added).

n14. See Remington Arms, 777 F. Supp. at 183 (citing Conn. Fund for the Env't v. Contract Plating Co., F. Supp. 1291, 1293 (D. Conn. 1986)).

n15. See Gwaltney, 484 U.S. at 60-61.

n16. See Remington Arms, 777 F. Supp. at 183-86.

n17. See *id.* (citing Conn. Fund for the Env't v. Contract Plating Co., 631 F. Supp. 1291, 1293 (D. Conn. 1986)).

n18. 33 U.S.C. 1365(a) (2000); see also Gwaltney, 484 U.S. at 53.

n19. 1319(d).

n20. 949 F.2d 552 (1st Cir. 1991).

n21. *Id.* at 526 (citing Gwaltney, 484 U.S. at 61).

n22. Friends of the Earth v. Laidlaw Envtl. Servs., Inc. (Laidlaw III), 528 U.S. 167, 168 (2000). In this Article, we cite to Laidlaw at the district, appellate and supreme court levels. For convenience, we have labeled the cases Laidlaw I (Friends of the Earth, Inc. v. Laidlaw Environmental Services (TOC), Inc., 956 F. Supp. 588 (D.S.C. 1997)), Laidlaw II (Friends of the Earth, Inc. v. Laidlaw Environmental Services (TOC), Inc., 149 F.3d 303, 305 (4th Cir. 1998)), and Laidlaw III (528 U.S. 167, 168 (2000)).

n23. Middlesex County Sewerage Auth. v. Nat'l Sea Clammers Ass'n, 453 U.S. 1, 21 (1981).

n24. 33 U.S.C. 1251(g) (2000).

n25. See, e.g., Metropolitan Sanitary Dist. of Greater Chicago v. U.S. Steel Corp., 332 N.E.2d 426, 432 (Ill. App. 1975).

n26. See, e.g., Laidlaw I, 956 F. Supp. at 607-10.

n27. See Water Quality Act of 1987, Pub. L. No. 100-4, 101 Stat. 7 (codified as amended in scattered sections of 30 U.S.C.). Among other things, the amendments amended Sections 1251, 1342, and 1344.

n28. EPA v. City of Green Forest, 921 F.2d 1394, 1398 (8th Cir. 1990) (quoting 33 U.S.C. 1251(a)).

n29. Laidlaw I, 956 F. Supp. at 588.

n30. 1365(b)(1)(B).

n31. 484 U.S. 49 (1987).

n32. *Id.* at 60.

n33. *Id.* at 54.

n34. *Id.* at 56.

n35. *Id.*

n36. *Id.* at 60.

n37. *Id.* at 61 (citing Water Pollution Control Legislation: Hearings Before the Subcomm. on Air and Water Pollution of the Senate Comm. of Public Works, 92d Cong. 707 ("Citizen suits ... are brought for the purpose of abating pollution." (statement of Sen. Eagleton)) (1971); H.R. Rep. No. 92-911, at 407 (1972) ("Citizens may institute suits against polluters for the purpose of halting that pollution." (additional views of Reps. Abzug & Rangel)).

n38. 523 U.S. 83 (1998).

n39. Pub. L. No. 99-499, 100 Stat. 1728 (codified as amended at 44 U.S.C. 11001-11050 (2000)).

n40. Steel, 523 U.S. at 88.

n41. *Id.*

n42. *Id.* at 109.
n43. *Id.* at 110.

n44. *Id.* at 102 (quoting Lujan v. Defenders of Wildlife, 504 U.S. 555, 560 (1992).

n45. *Id.* at 103-04 (citations omitted).

n46. *Id.* at 105.

n47. *Id.* at 105-06.

n48. See Laidlaw III, 528 U.S. at 181.

n49. *Id.* at 183 (citing Sierra Club v. Morton, 405 U.S. 727, 735 (1972)).

n50. 528 U.S. 167 (2000).

n51. The facts as laid out in this paragraph are from *id.* at 175-76.

n52. The facts in this paragraph come from Laidlaw I, 956 F. Supp. at 593-94.

n53. The facts as asserted in the following two paragraphs are from Laidlaw III, 528 U.S. at 177.

n54. *Id.*

n55. *Id.* at 167; see also 33 U.S.C. 1365(b)(1)(B) (2000).

n56. Laidlaw III, 528 U.S. at 177-78.

n57. Laidlaw I, 956 F. Supp. at 607-08.

n58. Laidlaw III, 528 U.S. at 178.

n59. *Id.*

n60. *Id.* at 179.

n61. *Id.*

n62. *Id.*

n63. *Id.* at 179.

n64. Laidlaw II, 149 F.3d. at 306-7.

n65. Laidlaw III, 528 U.S. at 167.

n66. *Id.* at 188.

n67. *Id.* at 195.

n68. *Id.* at 181.

n69. Laidlaw III, 528 U.S. at 167, 179.

n70. 42 U.S.C. 2000(d) (2000).

n71. Robert D. Bullard, Dumping in Dixie: Race, Class and Environmental Quality 35 (1990).

n72. *Id.* at 42.

n73. *Id.* at 37.

n74. *Id.* at 42.

n75. *Id.* at 40.

n76. See generally United Church of Christ, Commission for Racial Justice, Toxic Wastes and Race in the United States (1987); U.S. General Accounting Office, Siting of Hazardous Waste Landfills and Their Correlation with Racial and Economic Status of Surrounding Communities, GAO/RCED-83-168 (1983); U.S. Environmental Protection Agency, Environmental Equity, Reducing Risks for All Communities (1993); Marianne Lavelle & Marcia Coyle, Unequal Protection: The Racial Divide In Environmental Law A Special Investigation: The Federal Government, in Its Cleanup of Hazardous Sites and Its Pursuit of Polluters, Favors White Communities over Minority Communities Under Environmental Laws Meant to Provide Equal Protection for All Citizens, a National Law Journal Investigation Has Found, Nat'l L.J., Sept. 21, 1992, at S2.

n77. See generally United Church of Christ, *supra* note 76; U.S. General Accounting Office, *supra* note 76; Lavelle & Coyle, *supra* note 76.

n78. See Thomas S. Mullikin, Business and the Environment, Address at the Environmental Justice Symposium, Charleston, South Carolina, (Jan. 22, 1999), in 65 Vital Speeches 333 (1999).

n79. Pub. L. No. 94-580, 90 Stat. 2795 (codified as amended in scattered sections of 42 U.S.C.).

n80. Tom Mullikin, Environmental Solutions 7 (1995).

n81. Mullikin, *supra* note 78, at 334.

n82. *Id.*

n83. Robert W. Collin & Robin Morris Collin, Sustainability and Environmental Justice: Is the Future Clean and Black?, 31 Envt'l L. Rep. 10,968, 10,978 (2001).

n84. This is a model of first impression that was developed for the project in south central Los Angeles.

n85. See Mullikin, *supra* note 78, at 335-36.

Tom Mullikin

As a senior environmental attorney with the Charlotte law firm of Moore & Van Allen, PLLC, Tom leads the firm's Government, Policy and Regulatory Affairs Team. His practice focuses on corporate compliance, regulatory relations and legislative representation. Tom's career spans more than twenty years and includes key legislative staff roles, lobbying activities, extensive environmental legal representation and management of environmental, energy and healthcare issue campaigns for industry. He previously served as Chief Counsel and Vice President for Public Affairs to the largest environmental services company in the world.

He has served on state and congressional staffs as well as an advisor and campaign manager to senior members of Congress and U.S. Presidential candidates from both the Republican and Democratic parties.

Tom led a team of researchers and environmental experts on an expedition to Antarctica in late 2005 to study the effects of climate change on the polar regions. He produced a subsequent video documentary, entitled: *Climate Change: Global Problems, Global Solutions* which has received widespread acclaim in both business and environmental circles. In October 2006, he led a second team to Namibia, Africa, to follow up on the findings of the Antarctica expedition and further explore the impact of global climate change on the fragile Sub-Saharan environment.

Tom is the author of *Truck Stop Politics: Understanding the Emerging Force of Working Class America*. He has been widely published in both legal and mainstream periodicals, including *Campaigns and Elections Magazine, South Carolina Jurisprudence, Vital Speeches of the Day, UCLA Journal of Environmental Law and Policy and Georgetown International Environmental Law Review*, and has been quoted as an expert in both the law and the environment by *The Los Angeles Times, Newsday, The St. Petersburg (Fla.) Times, The Charlotte Observer, The Rocky Mountain News, The Salt Lake Tribune*, and *Huntinamibia*.

http://www.mvalaw.com
http://www.globalclimatesolution.com